T0185800

SpringerBriefs in Fire

Series Editor

James A. Milke, University of Maryland, College Park, MD, USA

More information about this series at http://www.springer.com/series/10476

Linda Makovicka Osvaldova

Wooden Façades and Fire Safety

Effects of Joint Type on Ignition Behaviour

 Springer

Linda Makovicka Osvaldova
Department of Fire Engineering
Faculty of Security Engineering
University of Žilina
Žilina, Slovakia

ISSN 2193-6595   ISSN 2193-6609   (electronic)
SpringerBriefs in Fire
ISBN 978-3-030-48882-6   ISBN 978-3-030-48883-3   (eBook)
https://doi.org/10.1007/978-3-030-48883-3

This Springer imprint is published by the registered company Springer Nature Switzerland AG
The registered company address is: Gewerbestrasse 11, 6330 Cham, Switzerland

This book is dedicated to my lovely family, colleagues, friends and all people.

Foreword

Throughout history, people have used wood and enjoyed its natural beauty, workability and practicality. As the World moves towards the use of low energy-intensive, low carbon-footprint materials, the natural inherent advantages of wood make it the sustainable material of choice. With this in mind, building designers are specifying wood products for a growing range of applications—for building structures, internal wall and ceiling linings, floor coverings through to building façades.

However, following recent devastating and tragic building façade fires (e.g. Grenfell Tower in London, Lacrosse building in Melbourne), attention has focussed on the use of external combustible cladding products on buildings. Worldwide, these fires have focussed attention on National building codes which have now been —or are in the process of being—reviewed, updated and amended where required to take account of the fire safety risk that façade materials can place on buildings. Not only does the façade material itself have an influence on ignitibility and fire spread, but also does the installation of the façade itself; with major considerations being the influence of fixings, façade support structure and joint details.

This publication, *Wooden Façades and Fire Safety—Effects of Joint Type on Ignition Behaviour*, presents the findings of an investigation into the fire behaviour of various jointing details in Spruce wooden façades and describes the findings of the testing program. This publication, and its investigations, add greatly to the body of knowledge regarding the safe design of wooden façades.

As part of a building design, building professionals need to consider all aspects in the fire safe design of buildings which includes the design and detailing of building façades. A wholistic fire safety building design will consider all aspects of the façade design with the influence of jointing system being an important consideration based on this investigation.

March 2020   Mr. Boris Iskra
                                            National Codes & Standards Manager

                                    Forest and Wood Products Australia Limited
                                                          Melbourne, Australia

Preface

Many research organizations deal with the issue of fire protection, especially combustible (ignitable) building materials. In addition to addressing the reaction to fire of materials and the fire resistance of structures, special research methods, procedures, experimental equipment and tests to monitor the specific conditions of fire generation and propagation are also modified. Such attention is also paid to facades. More-over to large-scale tests, there are a number of publications available dealing with structural models, verification of façade and cladding compositions. Their design and material composition are also verified.

A certain blank was dedicated to the influence of joints of wooden façade elements on the origin and spread of fire. The joints of these wooden elements logically appear to be the weakest point. It can accumulate heat, penetrate the joints itself and attack other materials in the structure. When air flow (outside or behind the structure) is added, there are hidden conditions for the creation and spread of fire. Today's technology and technology of wood processing allow to apply joints of various constructions to the facades, which can meet the demanding ideas of architects and have not only technical but also aesthetic significance.

Verification of the impact of joints on the possibility of their ignition as well as possible transmission of fire along the facade is supported by our research, which is presented in this publication. Methodology as well as laboratory equipment was developed using a low-value radiant and flame heat source. Evaluation criteria have also been proposed. The results measured by us show that the joint of the facade elements influence the possible origin and spread of fire along the facade.

This publication, *Wooden Façades and Fire Safety—Effects of Joint Type on Ignition Behaviour*, is devoted to wood processors, companies that deal with wooden constructions, or wood-based cladding and facades. Designers and architects, assessors and fire protection experts as well as researchers and students can find valuable information here.

Žilina, Slovakia
March 2020

Linda Makovicka Osvaldova

About This Book

This book, *Wooden Façades and Fire Safety—Effects of Joint Type on Ignition Behaviour*, contains new non-standardized values of the behavior of joints of wooden elements under thermal load that are part of wooden facades. Wood cladding is subject to numerous studies from several perspectives, i.e. technical/construction solution of facing structures, the evaluation of materials from the fire protection perspective, their certification as well as the fire regulations, which allow, or prohibit, the use of flammable materials such as wood. All of these are described in the international or national legislation. The missing information in these documents was the effect of joints (lengthwise in our case) of different wooden elements. Since wood can be joined in various ways (by means of an adhesive, metal, wooden connector or using no other auxiliary materials), different types of connections were chosen and subjected to the conditions of an emerging fire. These conditions are represented by a radiant heat source as well as a flame heat source of a low intensity and exposure for a short time.

The conditions of the experiment are set so that the joint type can manifest itself for selected evaluation criteria. The laboratory equipment is simple and set in an open environment simulating the conditions of a real fire. A great deal of attention was paid to the selection of the samples and to their moisture level and density, since these two physical values of wood significantly influence the relationship of wood to fire and ignition. Therefore the selection of wood with the proper density and moisture level was very rigorous so that the significance of the joint becomes apparent. Weight loss, burning rate, ratio a/b (see the publication) and the size of charred layers were the selected evaluation criteria. The monograph incorporates introductory chapters, which introduce the issue, describe the precise research methodology and present the results mainly graphically. The paper is intended for timber construction workers, construction companies that are dealing with wooden facades and cladding. All measurements as well as the methodology stated in the publication are original.

Contents

Acronyms and Abbreviations

BS	British standards
d0, d1, d2	Flaming droplets/particles in the test according to EN 13823
DIBT	German Institute for Construction Technology (Deutsches Institut für Bautechnik)
DIN	German institute for standardization (Deutsches Institut für Normung)
EN	European norm
EU	The European Union
F_s	Flame spread (EN 13501-1) (mm)
FIGRA	Fire Growth RAte index (EN 13823)
ISO	International Organization for Standardization
K	Type of thermocouples
LFS	Flame propagation in horizontal direction (EN 13823)
$m(0)$	Weight of the sample in time $(\tau 0)$ (g)
$m(600)$	Weight of the sample in time $(\tau 600)$ (g)
$m(48)$	Weight of the sample in time $(\tau 48)$ (g)
$m(\tau)$	Sample weight in time (τ) (g)
$m(\tau + \Delta\tau)$	Sample weight in time $(\tau + \Delta\tau)$ (g)
MDF	Medium-density fibreboard
OSB	Oriented strand board
ÖNORM	Austrian standards
P	Ratio of maximum burning rate divided by the time when it is reached
PSC	Combustion heat (MJ/kg)
s1, s2, s3	Smoke production (EN 13823)
S_C	Total surface (%)
S_{CH}	Charred surface (removed layer) (%)
S_P	Pyrolytic layer (light and dark brown layer) (%)
S_N	Layer not affected by fire (%)
SMOGRA	Smoke Growth RAte index (EN 13823)

STN EN ISO	Slovak technical standards
t_f	Sustained flaming (ISO 1182)
THR_{600s}	Total heat release from the sample during $300 \text{ s} \leq t \leq 900 \text{ s}$ (MJ)
v_r	Relative burning rate (%/s)
$v_{r\ max}$	Maximum burning rate (%/s)
WATT91	DIN EN 14257
$\delta m(\tau)$	Relative weight loss in time (τ) (%)
$\delta m(\tau + \Delta\tau)$	Relative weight loss in time ($\tau + \Delta\tau$) (%)
$\delta_{m\ 48}$	Difference in relative weight loss (%)
Δm	Weight difference (%)
Δm	Weight difference (g)
ΔT	Temperature interval (°C)
$\Delta\tau$	Time interval where the weights are subtracted (s)
τ	Time when maximum burning rate is achieved (s)

List of Figures

List of Tables

Chapter 1
Introduction

Abstract This chapter deals with the issue of façade fires of buildings in general. After presenting general information on the major causes of façade fires, it deals with possible solutions for this problem and includes potential scenarios of ignition and development of façade fires. We mainly pay attention to wooden façades. The chapter treats the issue of certifications of materials mainly on wooden façades and multi-story buildings. In addition to the basic tests used for certification of materials, we provide information on special tests for assessing façades. Design solution of façades is a subject of interest as well, along with fire safety regulations.

Keywords Fire · Building façades · Material solution of façades · Construction solution of façades · Legislation for façades · Fire protection

Wood. Wood might not be as rare and highly valued as other materials, however, it has many advantages. Unlike other raw materials, whose resources could be depleted within the next ten to one hundred years, wood is a renewable material and if people are economical with it, it might also be inexhaustible.

Wood has some technical characteristics predetermining it to be suitable for general use—furniture industry, musical instruments, various structures and artistic objects. Both positive and negative characteristics can be attributed to wood. Does wood have any negative characteristics at all? Yes, it does. The trunk can be shaped into many forms, but the dimensions of products are limited by the size of a tree; water continuously circulates in wood; it changes its dimensions depending on the humidity; it is prone to cracks, shrinking and many others defects; it is biologically degradable and it can burn. Getting to know these "flaws" and trying to eliminate them reflects how skilful people can be.

Deepening our knowledge of its internal structure, chemical composition, physical characteristics and mechanical properties also stems from the intensive technological progress regarding its processing and multilateral use. Wood is processable and workable in a rather easy and economical way. In addition to technical and aesthetic qualities, wood also creates a positive psychosomatic microclimate for people.

It is questionable whether wood combustion is a negative property. In order to use fire to our advantage, mankind certainly used wood as the best resource for fuel. The first and sole function of wood was that is was used as fuel. It is only later when it started to serve other purposes, e.g. building material and material for making various types of products or items of daily use.

Its flammability is suddenly becoming an undesirable phenomenon-wood catches and spreads fire. Wood in any form (raw material, semi-finished product, finished product) and processed in any way is a flammable material. Its ignition and burning, however, can be regulated in different ways. Modification of its parameters, in relation to fire, is needed, mainly in terms of wooden structures or buildings where applied. In wooden constructions, the formation and development of fire can be influenced and regulated in several ways, but it is not just the material which can pose problems. However, the material is the starting point. The "position" of wood within a structure must be determined since it might manifest itself in different ways during a fire. e.g. wooden cladding along which fire spreads very quickly. As noted in this Chapter, it is not always the wooden cladding, which might contribute to the spread of fire. Other materials applied onto the façade may be the cause as well. A great deal of attention has been paid to this issue in terms of material and construction perspective, certification, standards and other norms and regulations.

In all these domains, it is wood itself which is being assessed, not joints. Since façades cover large surfaces, it is necessary, on wooden facings in particular, to join the elements. Pieces of wood can be connected using various types of materials (glue, metal fasteners) as well as woodworking joints (see Chap. 2). And it is the effect of joints on a potential fire and its spread that has been the subject of our research.

To meet the research objectives, it was necessary to select the tree type (spruce wood was chosen for the experiment), laboratory equipment, evaluation criteria, as well as the types of joints. After the first experiments with a radiant heat source simulating a wooden façade facing a fire coming from a neighboring building, the flame source has been introduced into the experiment. The exposure time of both heat sources was an important criterion to get informative values of the evaluation criteria. Heat load needs to be accurate-neither too low so that is manifest itself in the results, nor too high-the sample might burn away completely. Detailed specifications of the experiment are described in Chap. 3. To set the limits and threshold conditions, it was necessary to get acquainted with spruce wood as such, its behavior in the process of burning as well as the effect of its properties i.e. density. At the same time, it was necessary to acquire knowledge of testing of wood and wood-based materials and test conditions using various certified tests to determine fire protection properties of materials. The testing was carried out only to obtain certain values, but the tests themselves have been subject to an objectivity evaluation study. Each test was aimed at the search for its specifics and the specifics of wood-based test specimens in order to achieve the highest objectivity possible. To avoid the general statement "wood burns" and to put an emphasis on the objectivity, we were looking for the conditions and physical parameters of the environment as well as the wood under which wood ignites and burns.

Chapter 4 presents both graphically and in tabular format the results obtained. The results prove that the parameters were set correctly, and, most importantly, that the idea to look into the issue of wood joints and their effect on the possibility of ignition and fire development on wooden facing was justified.

The information provided in this publication, *Wooden Façades and Fire Safety—Effects of Joint Type on Ignition Behaviour*, should help the designers of façade structures or wooden buildings, engineers, as well as the fire protection engineers. I believe that even research workers dealing with the issue of combustible materials can draw inspiration from this book.

1.1 Fire Spreading Along Façades

According to engineers and security experts, coating, which is in many cases made from highly flammable materials, is often to blame for the cause of fire [1]. Fires happened under similar circumstances in high-rise buildings. One of the largest fires occurred on January 1st at the Address Downtown, a luxury hotel in close proximity to the highest skyscraper in the world—Tamwel Tower, Dubai 16 people suffered injuries, mostly minor ones. The cladding was made from non-combustible materials and the building was equipped with high-quality fire equipment. A similar fire engulfed a high-rise building in Busan City, Korea. The fire in London had disastrous consequences, where fire spread along the façade.

1.1.1 Problem Solving

The application of flammable materials (e.g. wood) for cladding has been very popular with designers and architects lately [2]. Their projects and designs require further measures to be implemented to prevent a fire. The measures aim to prevent uncontrolled spread of fire along the surface or in openings. By implementing these measures, wood can be used without reducing the required security level concerning fire protection. These measures can be divided into different categories [3]:

- certification,
- material solution,
- design solution,
- legal solution in the field of fire protection.

Cladding design is a multidisciplinary process (Fig. 1.1). This multidisciplinary process aims to ensure technical and technological optimization to achieve the most optimal design of external cladding [1, 5, 6], on the basis of several works, compiled a study of fundamental criteria for external cladding optimization.

The optimization itself lies in a few basic criteria, such as sustainability [7–11], user friendliness of a building (thermal, acoustic and optical) [12–14], architectural and design solution [15] static resistance, fire safety and so on.

The problem lies in the additional thermal insulation of buildings using flammable materials as well as an increased demand for wood from aesthetic and functional point of view. In many legislations, special attention is paid to external wooden cladding

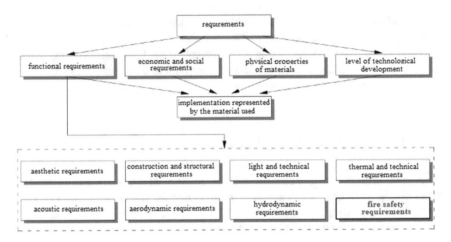

Fig. 1.1 Factors involved in external cladding fire [15]

where strict criteria and conditions for such cladding are implemented. In general, the legislation is based on fire scenarios concerning potential ignition and subsequent fire of façades, which are shown in Fig. 1.2 [2].

Scenario A: The fire spreads from the neighboring buildings (or other fire sources defined in the legislation) onto the building. The fire risk analysis is, for the majority of applications, represented by ignition risk assessment when the façade is exposed to a thermal radiation source. If the buildings are in close proximity to each other, the contact with the flame and the ignition of materials can be analyzed as well. This scenario also takes into consideration the spread of fire from a neighboring building onto the neighboring buildings. For flammable façades, it is necessary to take into consideration the heat from the burning façade—if the façade gets ignited—as well as radiant heat from openings and flames coming from the cavities on the façade of the neighboring building. The issue of the distance between the adjacent buildings is also included in the fire protection legislation for this scenario.

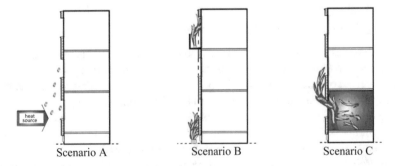

Fig. 1.2 Fire scenarios for façade fires [20]

Scenario B: The fire spreads from an external source adjoining the façade (not a neighboring building) e.g. a fire of a vehicle, litter bin, etc. including the balcony area (see the upper part of Fig. 1.2 scenario B).

Scenario C: A vertical fire spreads between the cavities from a fully developed fire inside a building with at least one cavity on the façade.

In the first stage (see Fig. 1.3), flames and hot gases damage internal surfaces of the room which are exposed to fire. In the early stages of fire development, the fire occurs exclusively inside the fire section. Fire is controlled by ventilation (excess of fuel, oxygen shortage). Depending on the size of fire section and the level of fire load, the flames are rising up as a result of pressure expansion and turbulent flow of flames onto the exterior surface of the external cladding [16]. At this stage, the fire is already controlled by fuel surface (oxygen surplus) and the risk of the fire spreading to the upper floors is increasing (step 2 Fig. 1.3) [5].

If, in this case, the external cladding contains flammable components, their presence in the structure might represent an additional thermal component, which, in addition to faster fire development, results in lower fire resistance. In the last stage, flames shooting out of open spaces are trying to ignite the combustible elements on the next floor through radiating elements of heat transfer to ensure a continuous dissemination of fire between the floors (step 3 Fig. 1.3).

Wood, a natural and renewable material, is becoming more and more popular when used in the exterior. Constructions implementing wooden façades, terraces and other outdoor wooden elements are growing in numbers. This trend has also caught on in the urban areas and is represented by wooden elements inside playgrounds, street furniture or small permanent and temporary structures and façade elements. In general, wooden façades became popular for their aesthetic and functional value. Wood as a construction material has a variety of advantages. It is characterized by low thermal conductivity, excellent resistance to atmospheric corrosion and ability to bind CO_2 in its mass. It is also easy to dress, handle, assemble, it is recyclable and, last but not least, it has positive effects on mental health.

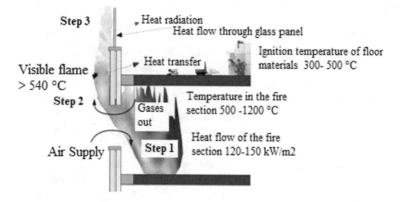

Fig. 1.3 Phases of fire development starting inside the building and spreading onto the cladding [5]

The application of wood for façade elements without increasing the fire risk requires perfect knowledge of its fire-technical properties and construction potential. Moreover, it is necessary to take into account all of the wood adjustments, including the ones which occurred in the production process, plus any other additional adjustments. Wood aging and the effect of climate and weather conditions on the quality of surface represent another very important type of "adjustments". Aesthetic, antifungal and insecticidal surface finish or fire-retardant application will also play a great role in the likelihood of a fire.

1.1.2 Certification

There is no doubt that testing all construction materials for fire protection purpose is necessary. The question is—how are we supposed to test it? The simplest or the most complex tests, both should simulate fire conditions to the last detail, so that it can predict the behavior of the material in case of fire. Incomprehensive tests may cause a biased evaluation of the material. This might mean two things, i.e. we will use the material which does not have adequate fire-technical properties, or we will not use a suitable material, because the test results were very unfavorable. It is essential to set the evaluation test criteria properly. One of the simplest as well as the most appropriate criterion is weight loss.

Test methods of materials for fire protection purpose can be divided into various point of views. The method—stage of the fire assessment—evaluates materials based on the first two stages of the fire, i.e. the likelihood of ignition—initiation and spread of fire. Special test method evaluating materials in spatial ignition stage—flash over. It is not necessary to assess fire resistance of the materials as such but to assess the construction as a whole representing the third stage of fire.

The second division is based on what the methods assess—solid substances such as construction materials, plastics, textiles, furniture and etc. Special methods—evaluating solid materials such as upholstered furniture, cables, clothing; waste materials such as various forms of dust in settled or whirled form; façade elements and constructions—can also fall into this category.

The last possible division of the test methods—division according to the size of the samples used for the test. For large samples, the effect of physical properties, adjustments or defects on the resulting value of the test can be taken into account.

The fire tests are sometimes combined with the other types of tests. They not only monitor basic physical processes of material burning, but also accompanying phenomena such as smoldering where the intensity and toxicity of the smoke are observed [3].

EU Standardization bodies aim to establish a harmonized procedure for reaction to fire classification system for building materials valid for all EU countries [17]. This classification is based on test procedures listed in the Table 1.1. The test results based on these methods are shown in Tables 1.2 and 1.3.

Table 1.1 Reaction-to-fire classification for construction materials excluding flooring [17]

The class	Test method	Classification criteria	Additional classification
A1	STN EN ISO 1182[a]	$\Delta T \leq 30$ °C	–
		$\Delta m \leq 50\%$	
		$t_f = 0$	
	STN ENI SO 1716	PSC ≤ 2.0 MJ/kg[a]	–
		PSC ≤ 2.0 MJ/kg[b, c]	
		PSC ≤ 2.0 MJ/kg[d]	
		PSC ≤ 2.0 MJ/kg[e]	
A2	STN EN ISO 1182[a]	$\Delta T \leq 50$ °C	–
		$\Delta m \leq 50\%$	
		$t_f = 20$	
	STN EN ISO 1716	PSC ≤ 3.0 MJ/kg[a]	–
		PSC ≤ 4.0 MJ/kg[b]	
		PSC ≤ 4.0 MJ/kg[d]	
		PSC ≤ 3.0 MJ/kg[e]	
	STN EN 13823	FIGRA ≤ 120 W/s	Smoke[f], burning droplets/particles[g]
		LFS < edge of the sample	
		THR$_{600s} \leq 7.5$ MJ	
B	STN EN 13823	FIGRA ≤ 120 W/s	Smoke[f], burning droplets/particles[g]
		LFS < edge of the sample	
		THR$_{600s} \leq 7.5$ MJ	
	STN EN ISO 11925-2[i] exposure = 30 s	$F_s \leq 150$ mm za 60 s	
C	STN EN 13823	FIGRA ≤ 250 W/s	Smoke[f], burning droplets/particles[g]
		LFS < edge of the sample	
		THR$_{600s} \leq 15$ MJ	
	STN EN ISO 11925-2[i] exposure = 30 s	$F_s \leq 150$ mm/60 s	
D	STN EN 13823	FIGRA ≤ 750 W/s	Smoke[f], burning droplets/particles[g]
	STN EN ISO 11925-2[i] exposure = 30 s	$F_s \leq 150$ mm/60 s	
E	STN EN ISO 11925-2[i] exposure = 15 s	$F_s \leq 150$ mm/20 s	Burning droplets/particles[g]

(continued)

Table 1.1   (continued)

The class	Test method	Classification criteria	Additional classification
F	without definition		

Explanatory notes for Table 1

[a]For homogeneous products and substantial components of non-homogeneous products

[b]For each external unsubstantial component of non-homogeneous products

[c]Alternatively, any external unsubstantial component having PCS 2,0 MJ/m^2 provided that the product meets the following EN 13823 criteria: FIGRA 20 W/s and LFS edge of the test sample and THR$_{600s}$ ≤ 4,0 MJ and s1and d0

[d]For each internal unsubstantial component of non-homogeneous products

[e]For the product as a whole

[f]In the last phase of the test method development, changes in the system of smoke detection that were introduced require further research. This may lead to a change of boundary values or parameters for smoke detection. s1 = SMOGRA 30 m^2/s^2 and TSP$_{600s}$ ≤ 50 m^2/s^2 = SMOGRA 180 m^2/s^2 and TSP$_{600s}$ 200 m^2/s^3 = fails to meet s1 or s2

[g]d0 = no flaming droplets/particles in the test according to EN 13823 during 600 s, d1 = no flaming droplets/particles persisting longer than 10 s for EN 13823 during 600 s, d2 = fails to meet d0 or d1

[h]Satisfactory = paper didn't ignite (no classification). Unsatisfactory = paper ignited (d2 classification)

[i]If the sample is exposed to flame and, where appropriate, from its end use perspective and when the edge of the sample is exposed to flame

Even though the basic material assessment is their reaction to fire (see Table 1.1), other testing methods assessing façade composition and structure were introduced. There is a whole range of middle-size and large-size tests. In this chapter, we will briefly introduce only the ones based on fire scenarios referred to in Fig. 1.2.

The method laid down in ISO 13785-1 [18]. Reaction to fire test for façades Part 1: The middle-size test determines the reaction to fire of the materials used for façades and cladding. It simulates external fire with flames reaching the façade (see Fig. 1.4). The method is not suitable for self-standing cladding and façades used as external façades for walls.

Test equipment consists of the sample, the support frame and a flame source. The frame holding the sample consists of three walls—back wall (3 parts) and two side walls. The rear side consists of mineral or stone wool—100 mm thick and at a density of 100 kg/m^3. The sample and the holder must be compact. The sample is placed 0.4 m from the bottom of the holder. The sample is mounted on the frame. All the measuring instruments are turned on two minutes before the test. The heat source is ignited. The test is photo documented or recorded (Fig. 1.5).

A propane burner of a rectangular shape of 1.2 × 0.1 × 0.15 m is used as a flame source. It is a flat-field burner having a special structure filled with crushed rock guaranteeing heat flow onto the test sample. 95% propane is used as a fuel. The sample represents a system we want to subject to testing and must be adjusted the way it is used in practice. It consists of cladding or façade panels covering the surface which is 1.2 m wide, 2.4 m high and 0.6 m wide and 2.4 m high. As we can see from Fig. 1.6, the sample consists of two parts, perpendicular to each other. The Fig. 1.4

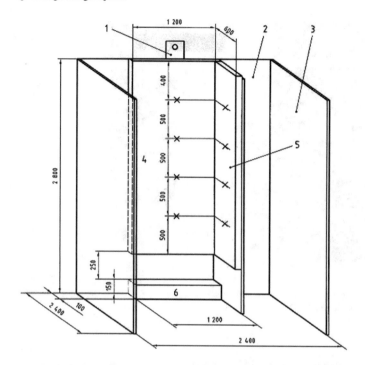

Fig. 1.4 Equipment [18]. 1—heat flow measurement, 2—back panel, 3—side wall (equipment), 4—back wall (sample), 5—side wall (sample), 6—burner, x—thermocouples

also shows the position of thermocouples (chromel/alumel, K type, with a diameter of 0.3 mm) on the sample. Temperature course recorded by thermocouples is evaluated and, at the same time, the behavior of materials during the test is described orally.

The Method laid down in ISO 13785-2. It specifies the test method for determining reaction to fire performance for materials and façade structures when exposed to heat and flame source of a simulated fire in the interior with flames shooting from the windows and stretching up to the façade (see Fig. 1.2 Scenario C). The data obtained from this test may also be applicable in the scenario of an external fire stretching up to a façade. This method is only applicable for façades and cladding that are not load-bearing. Determining the structural strength of the façade or the cladding is not needed.

The test is not designated for determining the fire behavior of the façade. Technical details of balconies, windows, curtains etc. are not taken into account in the test. This test does not include any risk assessment of fire spread, e.g. through window details of the façade system, because it is built only as a façade wall. The evidence proves that the inward corner (also called reentering corner) causes more intensive exposure to fire than a flat façade. The most frequent inward corner forms an angle of 90°. Test façade sample consists of inward corner of 90° (ISO 13785-2) [19].

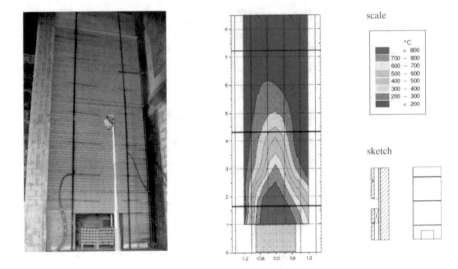

Fig. 1.5 Isotherms of maximum temperatures in front of façade surface of 5 mm and fire load of a 25 kg wooden batten [20]

The test can be performed in a combustion chamber in a room of least 20 m³ and a maximum of 100 m³. Samples must be designed the way they are used in practice. During the test, the temperature is measured using thermocouples (chromel/alumel, K type, diameter of 0.3 mm) and, at the same time, heat flow is measured by radiometers. The fuel may be in liquid form such as heptane or acetone and approximately 60 L are used in special burners or wooden cages of 450–500 kg/m³ with a 10–12% humidity rate.

A great number of large-size tests have been carried out testing e.g. façade composition, its structure etc. Several fire tests have shown that interior fires have the worst effect on façades [20]. This scenario causes the greatest energy release in front of the façade. When comparing the B and C fire scenarios, it is possible to determine conservative values for façade testing and evaluation:

- Energy release in front of the façade 1–1.5 MW
- Average length of flames 2.5–3.0 m (max. 6.0 m)
- Overall time of heat exposure 15–20 min
- Total time of flame exposure 10–15 min.

Figure 1.5 shows isotherms reaching peak temperatures in front of the façade with the distance of 5 mm and the fire load of a 25 kg wooden batten. Actual fires and fire tests show that all parts of the façade react to direct flame in the arc area depending on their flammability level. Windows or openings on the façade represent, in this area, weak points from fire protection perspective, whether they are open or closed.

Figure 1.5 shows the maximum temperatures during the fire test. The fire load in the test is based on temperature measurements in the tests of an actual fire.

1.1.3 Material Solution

The application of certified materials represents the basic material solution.

A detailed Table 1.2 with a possible application of wooden materials is indicated in [23]. It also indicates the end use conditions for the material.

Certification is essential. The application of non-certified materials in the buildings is currently very rare. The scientific community, however, is filling in the information which may be useful so as to improve the safety of buildings. There is a whole range of measurements treating the issue of wood selection and the conditions for its application. One of the articles [34] describes the effect of wood type—depending on the tree species type—on weight loss when exposed to thermal flame source of a low-intensity and short time exposure (using propane-butane burner from the laboratory equipment STN EN ISO 2592 [32]. Determining flash and fire points. Cleveland open-cup method. A gas burner was the only part of the equipment used. The sample was exposed to flame for 60 s only. Samples of $60 \times 20 \times 20$ mm (length x width x thickness) were used. Figure 1.6 shows that the effect of tree species type in constant conditions of the experiment is significant. This data was obtained from a scientific experiment, not from a certified test. Weight loss as well as the new evaluation criterion—weight loss 48 h after the experiment—were used to evaluate the experiment. This evaluation criterion will be applied in our experiment as well (see Sect. 4.4). It was essential to use the new evaluation criterion for all of the tree species which means that the wood was still burning flamelessly and higher weight loss values were observed than after the experiment. The marked values were recorded for soft deciduous woody plants such as alder, poplar or even maple [33].

In addition to wood selection, some adjustments need to be taken into consideration as well. Thermally modified wood or "thermowood" has been preferred recently. In many areas, positive characteristics of wood modified this way, in particular with regard to its biological resistance, are coming to the forefront. The fire-technical properties of such wood haven't been studied in detail yet. In the case of tropical tree species, even less information is available. Therefore, we have carried out an experiment using untreated as well as thermally modified teak wood with weight loss and relative burning rate being the evaluation criteria (see Figs. 1.7 and 1.8). The effect of thermal modification of teak wood is clear from the pictures. We would like to draw particular attention to the relative burning rate in the very first seconds of the experiment and the significant difference between untreated and thermally modified spruce wood (Fig. 1.8). Not only is the value higher but it was also attained earlier than with untreated wood [34].

Table 1.2 Classes of reaction to fire performance for wood-based panels [20]

Product	EN product standard	End use condition[6]	Min. density (kg/m³)	Min. thickness (mm)	Class (excl. floorings)
Cement bonded particleboard[a]	EN 634-2 [21]	without an air gap behind the panel	1000	10	B-s1, d0
Fibreboard, hard[a]	EN 622-2 [22]	without an air gap behind the wood-based panel	900	6	D-s2, d0
Fibreboard, hard[c]	EN 622-2 [22]	with a closed air gap not more than 22 mm behind the wood-based panel	900	6	D-s2, d2
Particleboard[a, b, e]	EN 312 [23]	without an air gap behind the wood-based panel	600	9	D-s2, d0
Fibreboard, hard and medium[a, b, e]	EN 622-2 [22] EN 622-3 [24]				
MDF[a, b, e]	EN 622-5 [25]				
OSB[a, b, e]	EN 300 [26]				
Plywood[a, b, e]	EN 636 [27]	"	400	9	D-s2, d0
Solid wood panel[a, b, e]	EN 13 353 [28]			12	
Flaxboard[a, b, e]	EN 15 197 [29]	"	450	15	D-s2, d0
Particleboard[c, e]	EN 312 [23]	with a closed or an open air gap not more than 22 mm behind the wood-based panel	600	9	D-s2, d2
Fibreboard, hard and medium[c, e]	EN 622-2 [22] EN 622-3 [24]				
MDF[c, e]	EN 622-5 [25]				
OSB[c, e]	EN300 [26]				
Plywood[c, e]	EN 636 [27]	"	400	9	D-s2, d2
Solid wood panel[c, e]	EN 13 353 [31]			12	
Particleboard[d, e]	EN 312 [26]	with a closed air gap not behind the wood-based panel	600	15	D-s2, d0
Fibreboard, medium[d, e]	EN 622-3 [27]				
MDF[d, e]	EN 622-5 [25]				
OSB[d, e]	EN 300 [26]				
Plywood[d, e]	EN 636 [30]	"	400	15	D-s2, d1

(continued)

Table 1.2 (continued)

Product	EN product standard	End use condition[6]	Min. density (kg/m^3)	Min. thickness (mm)	Class (excl. floorings)
Solid wood panel[d, e]	EN 13 353 [28]				D-s2, d0
Flaxboard[d, e]	EN 15 197 [29]	"	450	15	D-s2, d0
Particleboard[d, e]	EN 312 [23]	with an open air gap behind the wood-based panel	600	18	D-s2, d0
Fibreboard, medium[d, e]	EN 622-3 [24]				
MDF[d, e]	EN 622-5 [25]				
OSB[d, e]	EN 300 [26]				
Plywood[d, e]	EN 636 [37]		400	18	D-s2, d0
Solid wood panel[d, e]	EN 13 353 [28]	"			
Flaxboard[d, e]	EN 15 197 [29]	"	450	18	D-s2, d0
Particleboard[e]	EN 312 [23]	any	600	3	E
OSB[e]	EN 300 [26]	"	400	3	E
MDF[e]	EN 622-5 [25]	"	250	9	E
Plywood[e]	EN 636 [27]	"	400	3	E
Fibreboard, hard[e]	EN 622-2 [22]	"	900	3	E
Fibreboard, medium[e]	EN 622-3 [24]	"	400	9	E
Fibreboard, soft[e]	EN 622-4 [30]	"	250	9	E

Explanatory notes for Table 2

[a]Attached without any air gap directly against the class A1 or A2-s1, d0 products with a minimum density of 10 kg/m^3 or of at least class D-s2, products d2 with a minimum density of 400 kg/m^3

[b]If mounted directly on a wooden base, a substrate of cellulose insulation material of at least E class can be used

[c]Mounted behind the air gap. Rear side of the cavity must be at least of A2-s1 class, d0 products with a minimum density of 10 kg/m^3

[d]Mounted behind the air gap. Rear side of the cavity must be at least of D-s2 class, d2 products with a minimum density of 400 kg/m^3

[e]Veneer, phenolic and melamine panels are included in the class except flooring

[f]Vapor barrier with up to 0,4 mm thick and weighting up to 200 g/m^2 may be mounted between the wooden panel and the base if there are no gaps between them

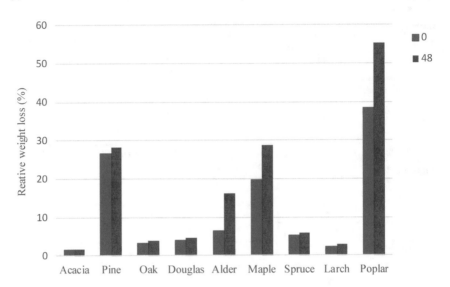

Fig. 1.6 Weight loss of selected tree species immediately after the experiment and 48 h after the experiment [33]

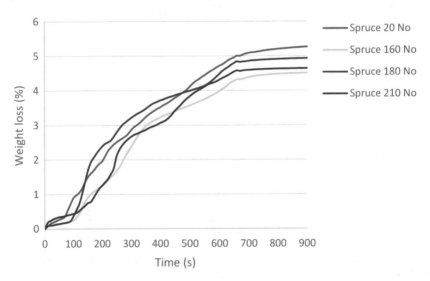

Fig. 1.7 Weight loss of thermally modified spruce wood [34]

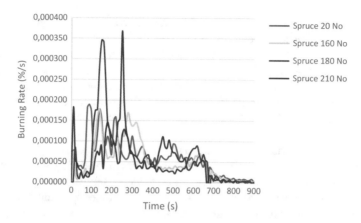

Fig. 1.8 Relative burning rate of thermally modified spruce wood [34]

Table 1.3 Classes of reaction to fire performance for wooden façade systems [20]

Façade systems Wood species	Wood species	Orientation	Joint width (mm)	Ventilation void (mm)	Insulation thick-ness (mm)	Class
Tongue/Groove-cladding, d = 19 mm	Spruce	Vertical		30		D-s1, 0
	Spruce	Vertical		100		D-s1, 0
Tongue/Groove-cladding, d = 9 mm	Larch	Horizontal	20	30		D-s2, 0
Tongue/Groove-cladding, d = 19 mm	Spruce	Horizontal	20	30		D-s2, 0
Tongue/Groove-cladding, d = 19 mm, battens 100 mm	Spruce	Horizontal		40	60	D-s2, 0
3-layered board, d = 19 mm	Spruce	–	20	30		D-s2, 0
3-layered board, d = 19 mm	Spruce	–		100		D-s2, 0

Similar experiments must be subject to certified tests so that the wood of each tree species, or any modification, are tested and certified for its safe application on a façade element or system as shown in Table 1.3. It also indicates the effect of the structure [20].

1.1.4 Design Solution

There are several design solutions for facade fire barriers. Some are simple i.e. wooden cladding treated with a fire-retarding agent. Wooden facade cladding has, in vertical direction, an aesthetic function. It is attached by means of hangers (hinges) at the bottom of the facade. The upper part is fixed by means of a thermally degradable substrate. On each side, a chain/cable is placed and it is attached a bit further from the edge of the cladding. The chain/cable is positioned so that it is not directly exposed to the fire. In the event of a fire, the upper linkage of the cladding thermally degrades and the wooden cladding of the facade switches, thanks to gravity, from vertical to a horizontal position (90°). This way, it delays the contact of a direct flame with a window above the fire.

In addition to material solution, design solution of the façade is important as well. An interesting experiment was carried out on a Suzuki model [35], where the effect of a barrier on temperatures on the upper floors was observed. Suzuki, in his work regarding a pilot study with a scale of 1:7, uses model isotherms to address the effect of barrier size ranging from zero to 250 mm. Calculations prove the importance of barriers to the fire spread, not only on the floor above the one where the fire originated but also on the upper floors.

The design solution contributes to the safety of facades, especially in respect of the fire spread along the facade and the transfer of fire to the upper floors.

Examples of the end use of cladding made of solid wood used for external cladding and combined with other types of materials, e.g. insulation or air gap behind the wooden cladding is shown in Fig. 1.9 [20]. Figures 1.10, 1.11, 1.12 and 1.13 show possible solutions of fire-fighting barriers (fire barriers in ventilation systems) [20].

Parameters of influence	Influence of fire behaviour		
	Best	Good	Critical
Type of cladding			

Fig. 1.9 Examples of influence parameters concerning fire behavior of multi-story façades [20]

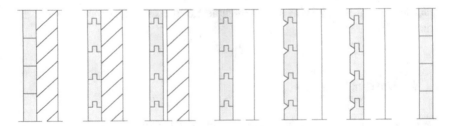

Fig. 1.10 Examples of end uses of solid wood paneling used as exterior wood cladding combined with another material (e.g. insulation) or an air gap [20]

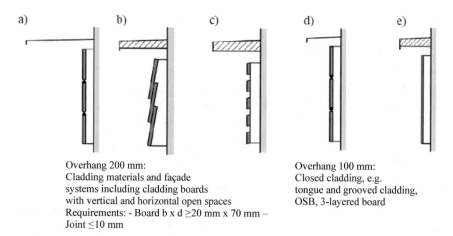

Overhang 200 mm:
Cladding materials and façade
systems including cladding boards
with vertical and horizontal open spaces
Requirements: - Board b x d ≥20 mm x 70 mm –
Joint ≤10 mm

Overhang 100 mm:
Closed cladding, e.g.
tongue and grooved cladding,
OSB, 3-layered board

Fig. 1.11 Firestop measures; **a** cantilever non-combustible, **b** timber or timber derived board and covering of non-combustible materials, **c** timber or timber derived board, **d** and **e** smaller design of firestops in comparison to a, b and c [20]

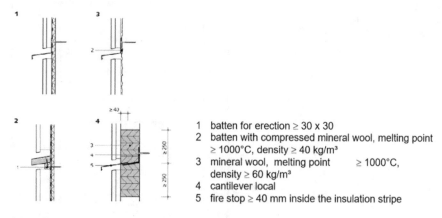

1 batten for erection ≥ 30 x 30
2 batten with compressed mineral wool, melting point ≥ 1000°C, density ≥ 40 kg/m³
3 mineral wool, melting point ≥ 1000°C, density ≥ 60 kg/m³
4 cantilever local
5 fire stop ≥ 40 mm inside the insulation stripe

Fig. 1.12 Firestop measures in rear ventilation voids [20]. 1—Screwed steel sheet apron, 2—Screwed timber board, 3—Screwed steel sheet apron on uneven surface, 4—Continuous apron on local cantilever

Joint coefficient values according to EN 1995-1-2 [38] is shown in Fig. 1.14.

Joint coefficient according to EN 1995-1-2 [38] applies to panel joints (cladding) not supported by structural elements or panels and to their effect on its protection performance (fire resistance). Joints thicker than 2 mm are not allowed under this regulation. Fire test results have shown that the effect of joints less than 2 mm thick is small. Simply put—the suggested method considers the effect of joints only for the last layer of the set on the side which is not exposed to fire and for the layer in front of an empty cavity. Special attention is paid to joints on wood-based façade cladding.

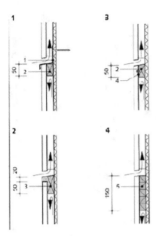

1 steel sheet thickness t ≥ 0,5 mm
2 timber batten, width ≥ 50 mm
   spacing of fasteners on battens ≤ 400 mm
3 timber batten grooved, width ≥ 70 mm
   spacing of fasteners on battens ≤ 400 mm
4 Leakproof connection between batten and cladding
5 Mineral wool
   Melting point ≥ 1000°C
   density ≥ 40 kg/m³
   width ≥ 150 mm, rear ventilation void leakproof filled
   fixing against slipping (e.g. screws)

Fig. 1.13 Firestop measures in rear ventilation voids [20]. 1—Timber batten with steel sheet, 2—Grooved timber batten, 3—Timber batten with leak proof connection to the cladding, 4—Mineral wool

Fig. 1.14 Joint coefficient k_j in the effect of joints in wood-based panels [38]

	type of joint	k_j
a	≤ 2 mm ►◄	0,2
b	≤ 2 mm ►◄ ◄ ► ≥ 30 mm	0,3
c	≤ 2 mm ►◄ ≥ 30	0,4
d	≤ 2 mm ►◄ ► 15 mm ◄	0,4
e	≤ 2 mm ►◄ ► ≥ 15 mm ◄	0,6

1.1.5 Legal Solutions in the Field of Fire Protection

Although the EU harmonized many regulations, there is a whole series of regulations and recommendations regarding the safety of wooden façades for individual EU member states. We will not deal with this issue in detail. We will just present a few examples—regulations directly stipulating the conditions for using wooden façades.

In Europe, fire risk classification for interior cladding significantly depends on a Single Burning Item Test (EN 13 823) [39] for the external façades. Additional criteria for multi-story buildings are applied in some countries. The approaches for multi-story buildings are mainly implemented. European countries require façade testing of multi-story buildings with flammable façades, as shown in Table 1.4 [46].

There are a number of regulations related to the spread of fire through the external façade which are important for wooden construction. The requirements concerning general fire resistance, which are met, are set out in an Australian regulation specifying the required types of constructions based on simple risk assessment that takes into account the buildings height and the purpose of the building. Bold hatching in Table 1.5 shows the types of buildings (number of floors), where wooden façades comply with the Australian regulations in force. A higher level of fire protection is necessary for multi- story buildings and buildings where the evacuation is likely to be slow [46].

The regulations (including Slovak regulations) limit the opening size to a maximum of one-third of the external wall surface on the floor where it is located. The regulation deals with external walls and cavities in the various fire sections within a single building and floor. It stipulates the distance between cavities depending on the angle between the neighboring walls, see Table 1.6. The bigger the angle between the walls is, the shorter the distance between the cavities, where an additional structure is needed, is. As soon as the wall reaches 180°, there are no limitations [46].

According to the requirements, the fire resistance of walls or ceilings depends on the floors, their dimensions and their end use. One or two family houses have less stricter requirements than multi-story buildings due to the number of occupants and shorter escape routes in buildings. To achieve the desired level of fire safety, it is necessary to take into account fire behavior of the building, service equipment and other safety measures depending on the given fire scenario. Evaluation criteria are interconnected and must be quantified i.e. different fire-resistant wall or ceiling sets, ignition period for flammable load-bearing structures (see Figs. 1.15) [20].

Table 1.4 Additional Fire performance requirements on Façades in some European Countries [46]

Country	Fire scenario	Test method	Exposure levels	Measurements	Façade test needed for wooden façades	Comments
UK	Flames out of a window	BS 8414 [40, 41]	15–75 kW/m^2 in 15–20 min	Damage, Heat flux, Temp	>2 storeys	
Switzerland	Flames out of a window	Large scale test	600–800 kW in 15–20 min	Damage, Temp	4–8 storeys	
Sweden	Flames out of a window	SP Fire 105 [42]	15–75 kW/m^2 in 15–20 min	Damage, Heat flux, Temp	>2 storeys	
Germany	Flames out of a window	DIN 4120-20 [43]	20–65 kW/m^2 (350400 kW) in 20 min	Flames, Glowing, Damage, Temp	≥4 storeys	Evaluation criteria to be decided by DIBT
Franc	Flames out of a window	Arête 10/09/197	15–75 kW/m^2 in 15–20 min	Flame spread, Damage, Temp	Depends on building type	Distance between buildings also to be respected
Austria	Flames out of a window	ÖNORM B3800 [44]	About 40 kW/m^2 in 20 min	Damage, Temp	4–5 storeys (ÖNORM B 3806)	Timber Constructions ÖNORM B 2332 [45]

From Fire Safety in Timber Buildings—Technical Guideline for Europe (according to [46] edited by the author)

Table 1.5 Type of construction required [46]

Building classification	Rise in storeys			
	1	2	3	4 and above
Class 2 Residential apartments	C	B	A	A
Class 3 Short–term accommodation	C	B	A	A
Class 3 Short–term accommodation	C	C	A	A
Class 5 Office	C	C	B	A
Class 6 Retail	C	C	B	A
Class 7 Carpark	C	C	B	A
Class 8 Factory or laboratory	C	C	B	A
Class 9a Healthcare	C	B	A	A
Class 9b Schools or assemble buildings	C	B	A	A
Class 9c Aged care	C	B	A	A
Class 9c Aged care	C	C	A	A

Table 1.6 Minimum distances allowed between openings in different fire compartments (Australian regulation) [46]

Angle between walls	Minimum distance (m)
0° (Walls opposite)	6
More than 0°–45°	5
More than 45°–90°	4
More than 90°–135°	3
More than 135° to less than 180°	2
180° or more	0

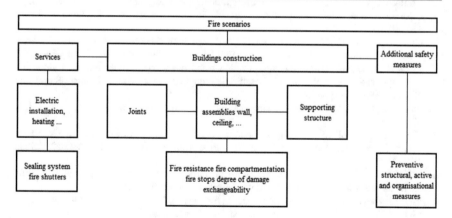

Fig. 1.15 Evaluation criteria for timber constructions [20]

References

1. Lesko R, Lopusniak M (2013) Regulative and standard requirements on façades in Slovakia and their mutual comparison with selected European countries. In: Proceedings of 2nd international seminar for fire safety of façades, Lund. EDP Sciences, pp 1–8. ISBN 978-1-5108-2406-5. Accessed 12 July 2017
2. Kotthoff I (2001) Brandschutz im Holzbau, 9. DGfH-Brandschutztagung, Würzburg Okt. (http://www.lignum.ch). Accessed 15 April 2017
3. Osvald A (2017) Wood fire protection. In: Fire protection, safety and security 2017, international scientific conference, conference proceedings, Zvolen. Publishing House of the Technical University in Zvolen, pp 193–200. ISBN 978-80-228-2957-1
4. Hidalgo-Medina JP (2015) Performance based methodology of the fire safety design of insulation materials in energy efficient buildings. Dissertation thesis, University of Edinburgh, p 429, (unpublished)
5. O'Connor D (2008) Building façade or fire safety façade? CTBUH J (2):30–39. www.jstor.org/stable/24192004. Accessed 26 March 2019
6. Hamdy M et al (2011) Applying a multi-objective optimization approach for design of low emission cost effective dwellings. Build Environ 46(1):109–123. ISSN: 0360-1323
7. Fesanghary M et al (2012) Design of low-emission and energy efficient residential buildings using a multi-objective optimization algorithm. Build Environ 49:245–250. ISSN: 0360-1323
8. Islam H et al (2015) Optimization approach of balancing life cycle cost and environmental impacts on residential building design. Energy Build 87:282–292. ISSN: 0378-7788
9. Asdrubali F et al (2013) Life cycle analysis in the construction sector: guiding the optimization of conventional Italian buildings. Energy Build 64:73–89. ISSN: 0378-7788
10. Damsky J, Gero J (1997) An evolutionary approach to generating constraint-based space layout topologies. CAAD Futures 1:855–864. ISBN: 0-7923-4726-9
11. Yu, W (2015) Application of multi-objective genetic algorithm to optimize energy efficiency and thermal comfort in building design. Energy Build 88:135–143. ISSN: 0378-7788
12. Croitoru C et al (2015) Thermal comfort models for indoor spaces and vehicles Current capabilities and future perspectives. Renew. Sustain. Energy Rev 44:304–318. ISSN: 1364-0321
13. Oral G K et al (2004) Building envelope design with the objective to ensure thermal, visual and acoustic comfort conditions. Build Environ 39(3):281–287. ISSN: 0360-1323
14. Li G, Hu H (2014) Risk design optimization using many objective evolutionary algorithm with application to performance based wind engineering of tall buildings. Struct Saf 48:1–14. ISSN: 0167-4730
15. Puskar A et al (2002) Building envelope-façades. Jaga group, Bratislava, p 338. ISBN: 80-88905-72-9
16. Kaihua L et al (2015) Merging behaviour of façade flames ejected from two windows of an under ventilated compartment fire. Proc Combus Inst 35(3):2615–2622. ISSN: 1540-7489
17. EN 13501-1: 2007 + A1: 2009 Fire classification of construction products and building elements. Classification using test data from reaction to fire tests
18. ISO 13 785-1: 2002 Reaction to fire tests for façades-Part 1: Intermediate-scale test
19. ISO 13 785-2: 2002 Reaction to fire tests for façades-Part 2: Large-scale test
20. Östman BL (2010) Fire safety in timber buildings. Technical guideline for Europe. SP report 2010: 19. ISBN 978-91-86319-60-1
21. EN 634-2 (2007) Cement-bonded particleboards-Specifications-Part 2: Requirements for OPC bonded particleboards for use in dry, humid and external conditions
22. EN 622-2 (2004) Fibreboards. Specifications. Requirements for hardboards
23. EN 312 (2010) Particleboards. Specifications
24. EN 622-3 (2004) Fibreboards. Specifications. Requirements for medium boards

25. EN 622-5 (2009) Fibreboards. Specifications. Requirements for dry process boards
26. EN 300 (2006) Oriented strand boards (OSB). Definitions, classification and specifications
27. EN 636: 2012 + A1 (2015) Plywood. Specifications
28. EN 13 353: 2008 + A1 (2011) Solid wood panels (SWP)
29. EN 15 197 (2007) Wood-based panels. Flax boards. Specifications
30. EN 622-4 (2009) Fibreboards. Specifications. Requirements for softboards
31. Makovicka Osvaldova L (2018) Density of test bodies and its effect on burning rate of fireretardant treated samples. In: 6th Asia conference on mechanical and materials engineering 1st edn. EDP Sciences, London, pp 1–4. https://www.matec/conferences.org/articles/matecconf/pdf/2018/72/matecconf_acmme2018_03002.pdf
32. STN EN ISO 2592 (2018) Determination of flash and fire points. Cleveland open cup method
33. Osvald A (2017) Weight loss as an assessment criterion for fire related properties. In. Advances in Fire and Safety Engineering 2017, Alumni Press Trnava, Trnava, pp 10–15. ISBN 978-80-8096-245-6
34. Osvald A, Gaff M (2017) Effect of thermal modification on flameless combustion of spruce wood. Wood Res 62(4):565–574. ISSN: 1336-4561, EV 3777/09
35. Suzuki T et al (2016) An experimental study of ejected flames of a highrise buildings–effects of depth of balcony on ejected flame. In: Proceedings of AOFST symposiums, pp 363–373. ISBN 978-981-10-0376-9
36. Schober KP, Matzinger I Brandschutztechnische Ausführung von Holz-fassaden, proHolz Austria (Hrsg.) Arbeitsheft 8/06
37. Lignum Dokumentation Brandschutz: 7.1 Außenwände, Konstruktion und Bekleidung (http://www.lignum.ch)
38. EN 1995-1-2 (2004) Eurocode 5. Design of timber structures. General. Structural fire design
39. EN 13 823:2012 + A1 (2016) Reaction to fire tests for building products-Building products excluding floorings exposed to the thermal attack by a single burning item
40. BS 8414-1 (2015) Fire performance of external cladding systems. Test method for nonloadbearing external cladding systems applied to the masonry face of a building
41. BS 8414-2 (2015) Fire performance of external cladding systems. Test method for nonloadbearing external cladding systems fixed to and supported by a structural steel frame
42. SP FIRE 105 (1985) Method for fire testing of façade materials, Dnr 171-79-360 Department of Fire Technology, Swedish National Testing and Research Institute
43. DIN 4102-20 (2017) Brandverhalten von Baustoffen und Bauteilen-Teil 20: Ergänzender Nachweis für die Beurteilung des Brandverhaltens von Außenwandbekleidungen
44. ÖNORM B 3800-1 (1988) Brandverhalten von Baustoffen und Bauteilen; Baustoffe: Anforderungen und Prüfungen
45. ÖNORM B 2330 (2007) Brandschutztechnische Ausführung von mehrgeschoßigen Holz- und Holzfertighäusern-Anforderungen und Ausführungsbeispiele
46. Alternative Solution Fire Compliance Façades. Technical Design Guide issued by Forest and Wood Products Australia. Prepared by: Exova Warrington fire Aus. Pty Ltd Suite 2002a, Level 20, 44 Market Street Sydney 2000 Australia First published: June 2013. ISBN 978-1-921763-68-7

Chapter 2
Joints

Abstract This Chapter deals with the possibilities of wood joining, with or without using additional joining elements (metal joints and adhesive). It emphasizes the wood which is to be jointed and its characteristics, beginning with its sawing method, drying method, to other technological adjustments so that the joint is firm enough and stable. We present wood joining methods from a historical perspective, as well as the latest technological methods using CNC machines. A great deal of emphasis is put on the quality of any joint, as joints may be the local source of fire development, in the event of a fire, because they may collapse.

Keywords Fire · Building façades · Wood joints · Adhesive · Metal joints · Longitudinal joints · Spatial joints · Joints in façades

The previous chapter treated the issue of claddings and façade materials and their design. In order to cover relatively large areas, it is necessary to join the individual fragments of the construction. We have therefore addressed this problem in a more detailed way. We focused on joining structural elements of natural spruce wood. Wood can be joined in multiple ways using various materials (glue, metal fasteners) or by using so-called woodworking joints.

2.1 Material

To assess the quality of each joint, it is necessary to have a good knowledge of wood characteristics. Besides its physical properties such as moisture and density (mentioned in the previous section), it is necessary to know the anisotropy of wood. Timber and its characteristics depend on the sawing method.

Besides the regular plain sawing, there are various special sawing methods (e.g. quarter sawing) or wood is sawn according to customer's requirements. These sawing methods (see Fig. 2.1) not only are more economical but they also facilitate wood processing and dressing. It is possible to make use of this fact when joining sawn timber, which can be used for cladding and façades and can be joined using carpentry joints. In the processing of coniferous raw material, the emphasis during handling

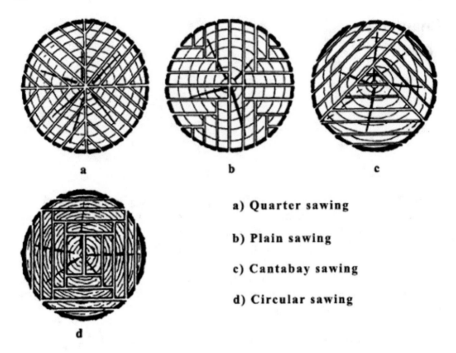

a) Quarter sawing

b) Plain sawing

c) Cantabay sawing

d) Circular sawing

Fig. 2.1 Special types of sawing methods [1]

is also placed, besides the dimensions and in relation to its surface structure, on the parts of a tree along the trunk:

- Bottom part
- Central part
- Upper part.

The sawing method and its quality become apparent during wood drying. Moisture loss brings about shriveling and shape changes in cut timber (see Fig. 2.1), which must be taken into account while joining the elements in longitudinal or crosswise direction. As shown in Fig. 2.2, if the orientation of annual rings is improper, it may cause the degradation of surface quality, greater material loss during machining, stress inside the glued joint and thus its weakening as well as other technological problems. These problems can also occur over time in the use of such structural element [2].

Fig. 2.2 Changes in shape due to the right way of joining timber crosswise

Fig. 2.3 Different mechanical properties of wood with regard to the direction of the cut

It is also important to get to know the mechanical properties of wood for different sawing methods shown in Fig. 2.3. If the tensile strength of the tongue on the left part of the Fig. is almost null in view of the angle of annual rings, the solution on the right side allows its practical use in view of the satisfactory mechanical properties and tensile strength of such a joint. Of course, this is the same tree species.

A careful examination of material then allows to create different wood joints either using other non-wood materials (glue and metal fasteners), or woodworking joints.

2.2 Design Solution

Woodworking joints are joints, in which the strength which is distributed is the contact pressure in a given joint and the friction inside this joint. These joints are made using wood exclusively. Mechanical elements are used only to fix the joint or to transfer additional forces. These joints have often been designed only based on the knowledge and experience of professional carpenters and were handed down from generation to generation.

Joints connecting wooden structures can be divided into:

- Mechanical fasteners (staples, coach bolts, screws, dowels, nails)
- Woodworking joints (scarf joint, cross lap joint, mortise and tenon, birds-mouth joint, etc.)
- Gluing (Glulam in particular, the so-called finger joint to create an infinite lamellae) [3].

Woodworking joints can be divided into:

- Adjusting joints (in linear direction)
- Coupling joints (in transverse direction)
- Panel point joints (wood joined at different angles and directions)
- Tenoned scarf joints.

Joints can be divided according to the nature and the type of the joining material:

- Flexible—woodworking joints
- Joints with mechanical fasteners
- Inflexible—glued joints [3].

Joints with mechanical fasteners can be divided according to the distribution of load inside the joint:

- Peg-like fasteners (staples, coach bolts, screws, dowels, bolt, nails)—bent and dented into the wood when the forces are transferred
- Surface fasteners (truss plates)—inserted or pressed into the wood and the forces are transferred on the surface of the components [3].

According to the position of structural elements and their purpose, there are three basic types of woodworking joints:

- Longitudinal joints (end-to-end butt joints, scarf joints) (Fig. 2.4) [3]
- Crosswise joints (mortise and tenon, lap joints, birds-mouth joints) (Fig. 2.5) [3]
- Spatial joints (strengthening joints, extending joints) (Fig. 2.6) [3].

Woodworking joints are quite complex, time consuming and require a great deal of precision. This was the reason why in the 1970's and 80's the use of woodworking joints significantly dropped. They made a comeback with CNC machines about twenty years ago. The machines enable to make such joints in a very precise, rapid and efficient way. This is the reason why these joints came into fashion again and began to be used to a greater extent. The joints in façade elements are not only quite firm and safe to use but they might also represent an aesthetic component in the façade.

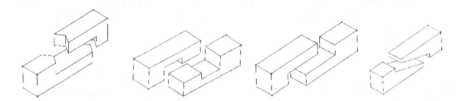

Fig. 2.4 Longitudinal joints [3]

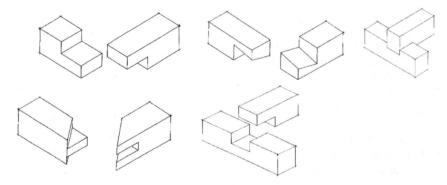

Fig. 2.5 Crosswise joints [3]

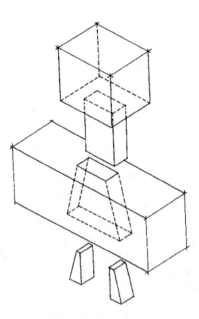

Fig. 2.6 Spatial joints [3]

To be able to use woodworking joints in a safe way, many engineering studies and experiments needed to be conducted, assessing the safety parameters of these joints. Specific dimensions, angles and bevel angles were set, ensuring the strength and safety of the given woodworking joint (see Fig. 2.7) [4].

Longitudinal joints are joints where a batten or a beam are lengthened. It can also be incorporated in façade elements. This lengthening can be carried out using a butt joint or a scarf joint. A butt joint is the weakest type of joint and it is necessary to

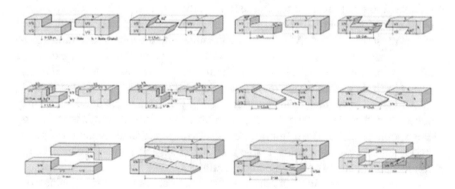

Fig. 2.7 Exact dimensions of each longitudinal woodworking joint [4]

brace it. A butt joint is a method of joining two members end to end. If the joint cannot be braced, we should use another type of joint such as the scarf joint.

A scarf joint is much stronger than a butt joint since it does not need to be braced. This joint is made as follows—we hollow out a part of a batten or beam which we are going to join and we make an inverse hole in the second piece of wood. By combining these gouges, we will get a compact piece of wood, which can be secured against movement in many different ways, i.e. by using nails in the diagonal of the gouge in the one-third distances. This joint may take different forms: half lap, beveled, dovetailed, tabled etc.

We only need a chainsaw or electrical saw to make such joint. When making such a joint, the most important thing is to keep the dimensions of sheets the same at both ends of the beams, which we want to connect. Joint can have various forms, depending on which material we are going to use to stabilize it since the joint itself is not stable on its own. We can either glue it or use different fasteners such as nails, screws, threaded bars with washers and nuts or truss plates [3].

The type of joining material affects the properties of the joint significantly. In addition, the position of the joining material is of crucial importance as well, mainly in the case of truss connector plates which could be placed on the side or on the top of the joint. Considering its properties, the main advantage is that it is easy to manufacture, it has good properties under load i.e. flexural strength. If we want to spread this joint, its properties are affected by the fasteners used. Screws and glue are better than wedges. Under torsional strength, this joint is not very resilient and quickly disintegrates [4].

It is becoming more and more popular to use adhesives and various glues. These joints are very firm but quite expensive. Some types of working joints are additionally glued (e.g. mortise and tenon joints). Glued joints are very tough, solid, flexible and waterproof. On the other hand, they require precise preparation and stronger pressures if pressed.

The adhesives are mostly synthetic thanks to their positive characteristics (water resistance, biological resistance) and user-friendliness. These adhesives are slowly replaced by animal glue and casein glue. Before gluing, wood must be planed, sanded and dust must be removed. After applying the glue onto the joint, the joint has to be pressed or clamped using stiffeners and clamps. Ideal compression pressure depends on the type of adhesive but in general, it should range between 0.4 to 1.2 MPa. The pressing itself then takes two or more hours depending on the type of adhesive [5].

The required strength is achieved after a few days. The ambient temperature may reduced as well as extend the time of gluing. Minimum ambient temperature during gluing should be 21 °C. The resulting strength also depends on the direction of grains of the beams. Ideally, the grains inside the joint should be perpendicular. It is not recommended to glue wood with parallel grains.

Different materials with different characteristics can be joined using adhesives. In joinery and furniture industry, various materials are glued together wood with plastic, rubber, textile [6]. The strength of a glued joint under normal conditions is always higher than the strength of wood alone.

Under thermal load as well as in the tests of fire-technical properties, can the adhesive layer become a better heat conductor compared with natural wood? A glued joint is definitely stronger than natural wood. In the tests of glued joints, it is the wood that disintegrates, the joint itself remains intact.

What impact does heating have? This has been studied by Osvald in his works [7]. The mechanical/strength properties of joints as well as their microscopic examination showed that a glued joint is stable even when under thermal stress. This means that from fire-prevention perspective natural wood can be replaced by glulam. This is evidenced by the microscopic images in Fig. 2.8, 2.9 and 2.10.

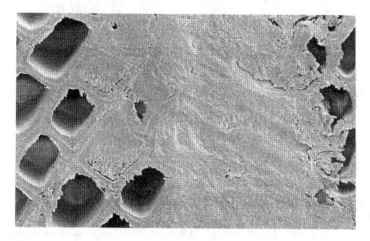

Fig. 2.8 Glued joint—intact wood structure, 15 mm below charred layer [7]

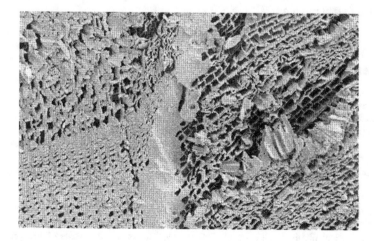

Fig. 2.9 Wood degradation—crack—along the glued joint, dark brown area and charred layer 7 mm below charred layer [7]

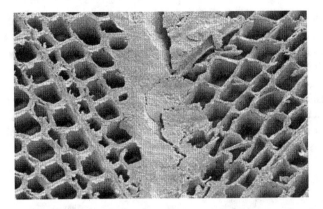

Fig. 2.10 Cracks in the wood, and in the glue 3 mm below charred layer [7]

As we can assume from the previous text, special attention is paid to woodworking joints. By improving machinery for their production, their use in practice is growing. As almost everything concerning wood—natural, inhomogeneous and anisotropic material—not even joints are a simple issue. Many improvements have been made to make them safe to use including their parameters and the regulations for their safe application. Their impact on the spread of fire remains unsolved (heat transfer through the joint, heating of the material behind the joint, etc.). Even when wood, in its various modifications, is tested for fire-technical properties, it is always tested in its compact form and the sample does not contain any joints. It is not possible to test the effect of joint using standardized test methods.

Therefore, an answer to the question "How can a joint of a wooden (= flammable) element influence the ignition and development of fire?' is this monograph which presents the results of experiments, where selected joints of spruce wood were subjected to thermal load.

References

1. Očkajová A, Kučerka M (2011) Materials and technologies 1. Wood technology. (Materály a technológie 1. Drevárske technológie). Matej Bel University in Banska Bystrica, p 115. ISBN 978-80-557-0262-9
2. Klement I, et al (2008) Basic characteristics of forest trees. National Forest Center. Accessed 17 April 2008
3. Reinprecht L, Štefko J (2000) Wooden cielings and roof constructions-types, faults, inspections and reconstructions (Dřevěné stropy a krovy-typy, poruchy, průzkumy a rekonstrukce) Praha. ABF, Nakladatelství ARCH, p 242. ISBN 80-86165-29-9
4. Sandanus J, Sógel K (2011) Timber structures–exercises and calculations, University script., Chapter No.7: Design of connections in a timber hall structure (in Slovak), pp 77–101
5. Osvald A et al (2009) Evaluation of material construction for the needs of fire protection. (Hodnotenie materiálov a konštrukcií pre potreby protipožiarnej ochrany). Technická univerzita vo Zvolene, Zvolen, p 335. ISBN 978-80-228-2039-4

6. Kupilík V (2012) Wooden buildings in the current low-energy construction in terms of fire safety (Drevostavby v súčasnej nízkoenergetickej výstavbe z hľadiska protipožiarnej bezpečnosti), [on line], TZB portál, ISSN 1338-3418. http://www.tzbportal.sk/stavebnictvo/drevostavby-v-sucasnej-nizkoenergetickej-vystavbe-z-hladis-ka-protipoziarnej-bezpecnosti. Accessed 3 April 2015
7. https://www.duvilax.sk/sk/drevarske-lepidla. Accessed 22 Mar 2019
8. Osvald A (2011) Wooden buildings ≠ Fire. (Drevostavba ≠ požiar). Technical University in Zvolen, Zvolen, p 336. ISBN 978-80-228-2220-6

Chapter 3
Experiment Description

Abstract This chapter describes the choice of materials and their classification according to density and laboratory conditions under which the selected joints of two wooden elements were tested, and which may be mounted onto the façade under the heat load of a radiant as well as flame heat source. In addition, this chapter describes the apparatus and the evaluation criteria, which indicate the suitability of the joint on a wooden façade.

Keywords Control sample—no joint · Glued joint · Dovetail joint · Biscuit joint vertical finger joint · Horizontal finger joint · Lap joint-screw with a nut · Lap joint-self-tapping screws

The following sub-chapters describe the experiment in detail including the goal of the experiment, material used, selected evaluation criteria and the description of the test joints and equipment.

3.1 Goal of the Experiment

Fire spreading along façades is affected by more factors than those mentioned in Chap. 1. The effect of these factors is evaluated at a number of levels and, in many cases, it is laid down in the standards or other regulations. Only a few study the effect of a horizontal joint in façade elements. This became the goal of our experiment.

3.2 Material Used for Experiment

Basic wooden material and some consumables were used for the experiment. In addition, glue and some auxiliary metal and wooden joining materials were used as well.

3.2.1 Wooden Material

Norway spruce (*Picea abies* (L) Karst Records.) [1] is the second most widespread wood in Slovakia and the most important one from the economic point of view. It is up to 50 m high columnar coniferous tree, its crown is evenly conical with the branches in whorls. Nowadays, spruce trees cover 24.92% of the forest area in Slovakia representing 481 466 ha (in 2012). Since 2005, the share of spruce trees has dropped by 1% [2].

Spruce wood is of yellowish to yellow-brown colour, very light, shiny, the pith has no colouring (see Fig. 3.1). It has very good load-bearing capacity. The lines of annual rings are definite, narrow, summer wood gradually turns into wide spring wood. There is a clear dividing line between the annual rings. The wood is light, soft, elastic, well-fissile, easy to stain and it is more difficult to impregnate it. It is free of defects and has typical symmetric and narrow annual growth rings (1–4 mm) with summerwood ratio in the annual ring ranging from 5 to 20%.

Some basic data on physical and mechanical characteristics of spruce wood [3]:

Physical characteristics:

- Density (at 12% moisture content) 441 kg/m^3
- Total longitudinal shrinkage 0.3%
- Total radial shrinkage 3.6%
- Total tangential shrinkage 7.8%
- Equilibrium moisture content (20 °C / 37% rel. humidity) 7.0% (20 °C / 83% rel. humidity) 16.4%.

Mechanical characteristics:

- Modulus of elasticity under bending 12500 N/mm^2

Fig. 3.1 Spruce wood (longitudinal, radial, tangential cut)

Table 3.1 Average density values for the given joint

Joint	Density
No joint—control sample	344.72
Glued joint	347.91
Dovetail joint	345.57
Biscuit joint	349.13
Horizontal finger	341.92
Vertical finger	342.28
Lap joint (screw with nut)	349.13
Lap joint (self-tapping screws)	341.33

- Modulus of rupture under bending 77 N/mm^2
- Tension strength 95 N/mm^2
- Compression strength 44 N/mm^2
- Brinell hardness perpendicular to the fibres 44 N/mm^2
- Janka Hardness 1.6 kN
- Nail withdrawal strength in N per mm depth and mm diameter 5.5 N/mm^2.

Before describing the test samples in detail, we will resolve the problem of wood selection the samples were made from. As a matter of course, humidity rate is $8 \pm 2\%$, but as several authors state [2] density has a large impact on the process of wood ignition and burning. Therefore, we paid a great deal of attention to this physical property of spruce wood when selecting wood for the samples. Density ranged from 340 ± 5 kg/m^3 (see Table 3.1). The average density values of the samples for the given joint are stated in the table. Here are the average values of 15 samples. This readout relates to wood only. Glued joints, biscuit joints, or joints with metal fasteners are not figured in this density readout. Samples used for measuring density had the dimensions of $160 \times 60 \times 20$ mm (length, width, thickness). The shape of the joint has been adjusted. After connecting and stabilizing it, the final dimensions were $250 \times 60 \times 20$ mm. Finally, a vibration grinder was used as the last adjustment method of the samples. 15 samples for each joint type were used for the experiment.

3.2.2   Consumables

Glue, biscuit, metal material, self-tapping screws, screws and nuts represented the consumables. The list of the consumables used for joints is stated in Table 3.2.

Table 3.2 Consumables used for the joint

Joint	Consumables
No joint—control sample	None
Glued joint	Adhesive
Dovetail joint	Adhesive
Biscuit joint	Adhesive
Horizontal finger	Adhesive
Vertical finger	Adhesive
Lap joint (screw with nut)	Metal
Lap joint (self-tapping screws)	Metal

Glued joint.

Würth Group glue was used as adhesive. It is highly resistant to heat and water. These qualities have been tested in accordance with the DIN EN 204 [4] norm in Institut für Fenstertechnik e.V. in Rosenheim.

The same institute confirmed that the test was in compliance with the DIN EN 14257 [5] norm (WATT91). The adhesive does not need to be mixed with any hardening agent and no processing time needs to be kept. After hardening, the glue is transparent, free of visible glued joints [6]. Joints are viscous-plastic after hardening. More details on this glue are available in the Table 3.3.

Table 3.3 Information on the glue

Appearance	Liquid
Color	White
Smell	Very weak, such as hydrocarbons
pH	3 concentrate
Melting/freezing point	Approximately 0 °C
Initial boiling temperature and distillation range	Approximately 100 °C
Flash-point	>100 °C
	Other information: it does not burn
Evaporation rate	Not available
Flammability (solid substance, gas)	Not applicable
Upper explosive limit/ upper limit of flammability	Not available
Lower explosive limit/ Lower flammability limit	Not available
Solubility (solubility in water)	Super-miscible
Auto-ignition temperature	Not available
Decomposition temperature	Not available
Explosive properties	Not explosive
Auto-flammability	Not pyrophoric

Fig. 3.2 Biscuit inside the
material and the biscuit itself

Biscuit joint

An industrially produced biscuit joint of $56 \times 23 \times 4$ mm was used. A biscuit joint was used according to the manufacturer's recommendations (see Fig. 3.2) The samples were milled out in the middle of its thickness by a biscuit joiner. A biscuit joint together with the gap and the entire surface of the joint were coated with adhesive.

A biscuit joint was made from a different material than the sample—i.e. beech plywood. Samples have been pressed in a press machine. Once the glue had hardened and dried, the samples were taken out from the press machine. Consequently, they were sawn and planed to a specific length and width.

Metal fasteners

Samples were made in the same way as the glued lap joints but they have not been glued with any adhesive and their dimensions were adjusted (sawed, planed, milled) before they were joined together. Instead of gluing, a coach bolt with a metric thread and a 5 mm thick, 30 mm long mushroom-head was used.

The screws were anchored with a washer and a nut on the other side. Before placing the bolts, the samples had to be drilled through with a drill of the same thickness i.e. 5 mm. Two of these screws were placed on each sample so that their centre is situated in the middle of the joint's width and in one third of the sample (see Fig. 3.9).

Samples were prepared the same way as the samples joined with a screw, yet they have not been pre-drilled completely—one part of the joint was drilled through with a drill which was 4 mm thick. These samples have been joined together with a galvanized, self-tapping, countersunk, 4 mm thick, 20 mm long head screw. These screws are placed in the same position as in the case of the lap joined samples joined with a screw and a nut (see Fig. 3.10).

3.3 Joints

All the joints, fragments of the joints, parts which were jointed are listed in Figs. 3.3, 3.4, 3.5, 3.6, 3.7, 3.8, 3.9 and 3.10 (width and thickness plan view).

Fig. 3.3 Jointless
sample—control sample [7]

Fig. 3.4 Lap joint—glued
joint [7]

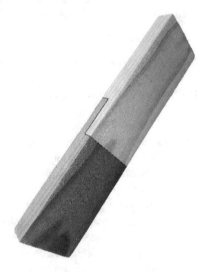

The following types of joints have been tested:

- Control sample—no joint (see Fig. 3.3)
- Glued joint (see Fig. 3.4)
- Dovetail joint (see Fig. 3.5)
- Biscuit joint (see Fig. 3.6)
- Vertical finger joint (see Fig. 3.7)
- Horizontal finger joint (see Fig. 3.8)
- Lap joint—screw with a nut (see Fig. 3.9)
- Lap joint—self-tapping screws (see Fig. 3.10).

Fig. 3.5 Dovetail joint [7]

Fig. 3.6 Biscuit joint [7]

Fig. 3.7 Vertical finger joint [7]

Fig. 3.8 Horizontal finger joint [7]

Fig. 3.9 Lap joint—screws
with nuts [7]

Fig. 3.10 Lap joint—with
self tapping screws [7]

3.4 Apparatus

A (non-standardized) apparatus was used for the experiment. The goal was to have a radiant heat source of rather low intensity in order to monitor the effect of joints on thermal degradation of samples by means of the selected evaluation criteria. The apparatus should make it possible to simulate the scenario A conditions (see Fig. 1.5, Chap. 1). Ceramic thermal radiator, shown in Fig. 3.11, was used for the experiment. Its maximum power output is 60 kW/m^2, 230 V and dimensions are 254 × 60 × 35 mm [9]. The distance between the sample and the thermal radiator is 35 mm. The sample was exposed to the heat from the radiator for 10 min.

The device is shown in Fig. 3.12.

Figure 3.13 is a schematic diagram of the thermal loading device testing the joints exposed to a flame heat source. The apparatus and its components are described in detail below the Fig. 3.13. Both apparatuses (for radiant and flame heat source) were

Fig. 3.11 Thermal radiator—radiation heat source

Fig. 3.12 The apparatus [8]

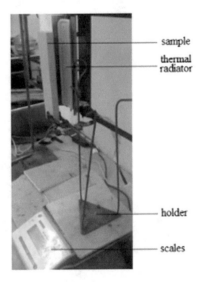

sample

thermal
radiator

holder

scales

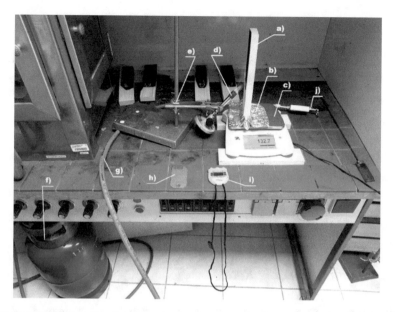

Fig. 3.13 Test equipment: **a** sample; **b** sample holder; **c** weight scales; **d** gas burner; **e** burner stand; **f** propane-butane cylinder; **g** gas intake; **h** (flame) measuring scale; **i** stopwatch; **j** data recorder [7]

designed so as to be open to atmospheric oxygen as well as for natural exhaust gas to be taken away. The intention was to simulate a real fire at its initial stage.

The Experiment Proceeded as Follows

- The radiator was warming up for 10 min to reach operating temperature
- The sample was mounted into the holder, the measuring process was initiated
- After 10 min, the holder was shifted from radiator so that spontaneous combustion didn't stop
- After 48 h, the samples was reweighed
- After 14 days the samples were cut in two through the middle

In case of flame heat source, the procedure remained the same.

3.5 Assessment Criteria

Assessment Criteria:

- Weight loss
- Relative burning rate
- *P* ratio
- Difference in weight loss 0–48
- Burned area and its size.

3.5.1 Weight Loss

When the samples were exposed to heat, we observed and recorded weight loss in 10 s intervals. Relative weight loss was calculated according to the relation (3.1):

$$\delta_m(\tau) = \frac{\Delta m}{m(\tau)} \cdot 100 = \frac{m(\tau) - m(\tau + \Delta\tau)}{m(\tau)} \cdot 100(\%) \tag{3.1}$$

where:

$\delta_m(\tau)$—relative weight loss in time (τ) (%);
$m(\tau)$—sample weight in time (τ) (g);
$m(\tau + \Delta\tau)$—sample weight in time ($\tau + \Delta\tau$) (g);
Δm—weight difference (g).

3.5.2 Relative Burning Rate

Relative burning rate has been determined according to the following relations (3.2) and (3.3):

$$v_r = \left| \frac{\partial \delta_m}{\delta\tau} \right| (\% \ /s) \tag{3.2}$$

or numerically

$$v_r = \frac{|\delta_m(\tau) - \delta_m(\tau + \Delta\tau)|}{\Delta\tau} (\% \ /s) \tag{3.3}$$

where:

v_r—relative burning rate (%/s);
$\delta m(\tau)$—relative weight loss in time (τ) (%);
$\delta m(\tau + \Delta\tau)$—relative weight loss in time ($\tau + \Delta\tau$) (%);
$\Delta\tau$—time interval where the weights are subtracted (s).

3.5.3 P Ratio

P ratio represents a maximum value of relative burning rate divided by the time when this value was reached. The ratio has been determined according to the following relation (3.4).

$$P = \frac{v_{rmax}}{\tau} (\% \ /s^2) \tag{3.4}$$

where:

P—ratio of maximum burning rate divided by the time when it is reached;
v_{rmax}—maximum burning rate (%/s);
τ—time when maximum burning rate is achieved (s).

Numerical value of this evaluation criterion gives relevant information on the behavior of the material during the experiment. If the maximum relative burning rate value is achieved in a short period of time, it is a negative sign. If the maximum value is reached later, it is a more positive indicator. It was necessary to compare these two indicators so as to emphasize the synergistic effect of the two measured values. An illustrative example is as follows: if the maximum burning rate value 10 is reached in the tenth minute, the monitored ratio reaches the value of 1. If it is achieved in the fifth minute, the ratio is 2; if in the first minute, it is 10.

3.5.4 Difference in Weight Loss 0–48

When exposed to the heat of the thermal radiator, weight loss is observed and recorded in several steps. Since the proposed apparatus as well as the overall methodology of the experiment proved itself to be sufficiently sensitive to the selected evaluation criteria, we decided to measure the weight of the samples after 48 h following the experiment. The weight is measured continuously for 600 s. Then the "final" weight loss under thermal stress is calculated (in 600-second). The weight loss after 48 h was then deducted from this value (Eq. 3.5).

This difference could also reach negative values for certain types of joints. This means that even though a higher weight loss was observed during combustion, it could be caused by thermal degradation as well as by drying of the wooden sample. If no intense burning of the sample occurred, it did not acclimate after 48 h and absorbed the moisture from the environment it was stored in, which resulted in a reduction of weight loss.

If the sample was burning flamelessly after being removed from the heat, it resulted in weight loss reduction, even if the sample could possibly absorb some moisture later on. It was not possible to monitor mass ratio by the effect of moisture and thermal degradation in the given experiment even if the samples had been placed in an air conditioned laboratory room. This evaluation criterion, as it was later proved, has some information value.

$$\delta_{m48} = \left(\frac{m(0) - m(600)}{m(0)} \right) - \left(\frac{m(0) - m(48)}{m(0)} \right).100(\%) \qquad (3.5)$$

where:

$\delta_{m\,48}$—difference in relative weight loss (%);
$m(0)$—weight of the sample in time ($\tau 0$) (g);
$m(600)$—weight of the sample in time ($\tau 600$) (g);
$m(48)$—weight of the sample in time ($\tau 48$) (g).

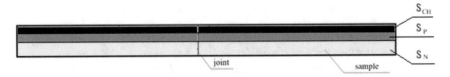

Fig. 3.14 Idealized zones of measurement

3.5.5 Burned Area And Its Size

The size of the burned area is measured after cutting the sample in two through the middle of its length. The charred layer is "removed" from the surface and the line of pyrolysis layer is determined. If the total surface of the sample is 5000 mm^2 (100%) before the experiment, the remaining intact area of wood and pyrolysis layer are measured out for each sample. Their surface is calculated, and the percentage of the original surface is determined. It is calculated according to the formula (Eq. 3.6). The idealized zone of measurement is shown in Fig. 3.14. The diagram shows the idealized measurement, linear borders of the measured surfaces. In fact, it is an uneven (bordered) area, which is measured directly on the samples.

$$S_C = S_{CH} + S_P + S_N (\%)$$
(3.6)

where:
S_C—total surface (%);
S_{CH}—charred surface (removed layer) (%);
S_P—pyrolytic layer (light and dark brown layer) (%);
S_N—layer not affected by fire (%).

References

1. http://www.tuzvo.sk/sk/organizacna_struktura/lesnicka_fakulta/oacnebreakdownof/depart ment/institute-forest-miningnobleif/contact/spruce.html. Accessed 22 June 2015
2. Klement I (2008) Swedes find worlds oldest tree [online]. news.bbc.co.uk. Accessed 17 Mar 2008
3. Klement I et al (2008) Basic characteristics of forest trees. National Forest Center. Accessed 17 Apr 2008
4. DIN EN 204 (2016) Classification of thermoplastic wood adhesives for non-structural applica- tions
5. DIN EN 14175-2 (2008) Fume cupboards-Part 2: Safety and performance requirements
6. https://www.duvilax.sk/sk/drevarske-lepidla. Accessed 22 Mar 2019
7. Benicky M (2019) Testing of selected carpentry joints with a flame source. Diploma Thesis. University of Zilina, Slovakia (unpublished)
8. Gabor T (2018) Testing of selected wooden joints for heat source reaction. Diploma Thesis. University of Zilina, Slovakia (unpublished)
9. http://hotset.sk/index.php?page=infraziarice. Accessed 17 Feb 2019

Chapter 4
Results of The Experiment

Abstract This chapter presents the results of measurements of selected joints under the heat load of radiant and flame heat sources. Results are presented graphically and in tabular form with a brief commentary. We state and comment on the average values of the measurements of the selected evaluation criteria, which predict the behavior of the joint in the case of a potential fire and its development.

Keywords Radiant heat source · Flame source · Weight loss · Relative burning rate · Ratio P · Weight loss difference 0–48 · Charred layer

The following sub-chapters describe the results of the experiment in detail including the partial evaluation for each evaluation criterion. Each sub-chapter deals with one evaluation criterion detailed in Sects. 4.1–4.5. The chapters are also divided according to the heat source type used for the experiment (radiant and flame heat source). It is not our aim to compare the two sources with each other and therefore they are evaluated separately. The aim is to compare the behavior of the selected joints and their reaction to short-term exposure to the two heat sources separately and to find out the properties of the joint in case of an emerging fire. The results are presented mainly graphically or in tables.

4.1 Weight Loss

The results of this evaluation criterion are presented in sub Sect. 4.1.1 for radiant heat source (Figs. 4.1, 4.2, 4.3, 4.4, 4.5, 4.6, 4.7, 4.8, 4.9, and 4.10) and in sub Sect. 4.1.2 for flame heat source (Figs. 4.11, 4.12, 4.13, 4.14, 4.15, 4.16, 4.17, 4.18, 4.19, and 4.20).

© The Author(s), under exclusive license to Springer Nature Switzerland AG 2020
L. Makovicka Osvaldova, *Wooden Façades and Fire Safety*,
SpringerBriefs in Fire, https://doi.org/10.1007/978-3-030-48883-3_4

4.1.1 *Radiant Heat Source*

As Fig. 4.1 shows, each joint type has an impact on the loss of weight under radiant heat source—the weight increases with each joint type in comparison with the control (jointless) sample. Figure 4.2—a bar graph shows the average maximum weight loss values for individual joints.

Figure 4.3 are the differences in maximum weight loss values for the individual joints after deducting the maximum weight loss of the control (jointless) sample. Based on these values we can assume that the glued joint, horizontal finger joint and lap joint with a self-tapping screw show the lowest weight loss values. Higher values are observed in case of screw with nut joint since the metallic element goes through wood in its entirety and tends to warm up faster as it has better thermal conductivity.

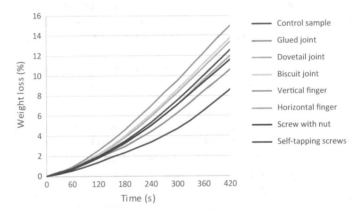

Fig. 4.1 Weight loss course of the given joints using radiant heat source

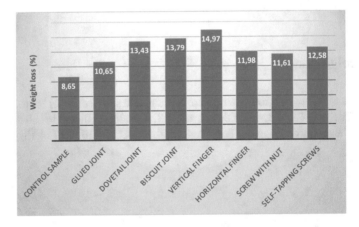

Fig. 4.2 Average weight loss values of the given joints using radiant heat source

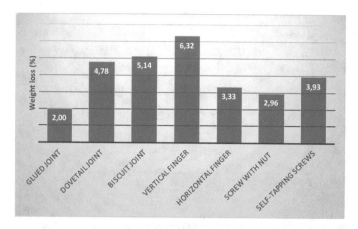

Fig. 4.3 Average differences in weight loss between a control sample and a joint using radiant heat source

This way it heats up the wood it is in contact with faster compared to lap joint with a self-tapping screw—self-tapping screw with nut is shorter, and therefore does not pass through the sample in its full width.

Significantly worse results we achieved for the dovetail joint, biscuit joint and mainly for vertical finger joint. Dovetail joint and vertical finger joint are designed the way allowing the heat to be transferred directly to the other side of the sample. The length of the indentation of the joint is 160 mm for dovetail joint and 360 mm for vertical finger joint. This certainly affects the weight loss of the given joint, even though the entire joint is glued.

Considering the joint's thickness, there is no obstacle to heat transfer compared to other joint types. As for the biscuit joint, there is a direct obstacle represented by the biscuit itself.

Higher weight loss values might be caused by the biscuit itself. The biscuit is made of beech wood which can thermally decompose faster than spruce wood mainly because of its higher hemicellulose content which tends to decompose faster at lower temperatures. Since both the joint and the biscuit itself were exposed to radiant heat source, more significant weight loss could be observed.

More details are provided in Figs. 4.4, 4.5, 4.6, 4.7, 4.8, 4.9, and 4.10. Figure 4.1 shows the weight loss courses for all the joints. Figures 4.4, 4.5, 4.6, 4.7, 4.8, 4.9, and 4.10 show only two curves—the test measurement, the jointless sample and the curve of the given joint.

Based on these weight loss curves, we can see how great the differences are and therefore they have been included in the monograph. As the photos prove, the joint has an impact on the burning rate course which, in this case, equals the weight loss.

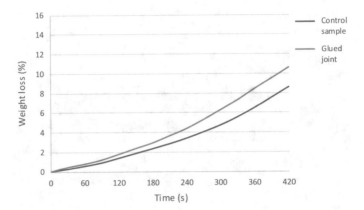

Fig. 4.4   Weight loss—control sample against glued joint using radiant heat source

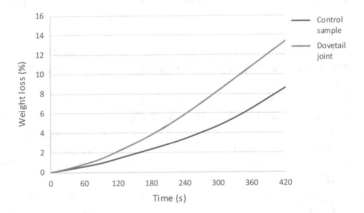

Fig. 4.5   Weight loss—control sample against dovetail joint using radiant source

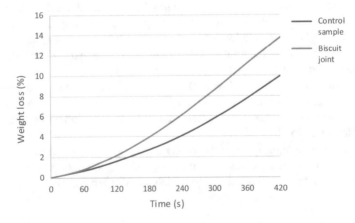

Fig. 4.6   Weight loss—control sample against biscuit joint using radiant heat source

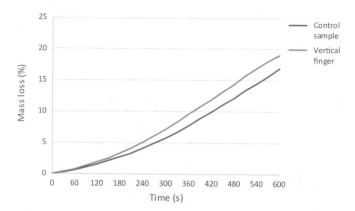

Fig. 4.7 Weight loss—control sample against vertical finger joint using radiant heat source

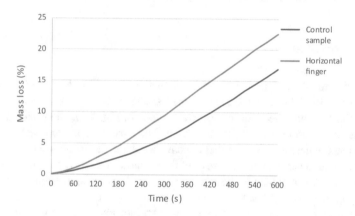

Fig. 4.8 Weight loss—control sample against horizontal finger joint using radiant heat source

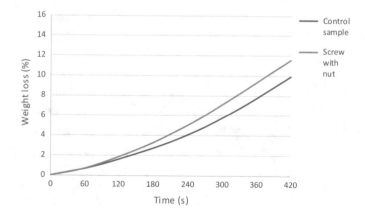

Fig. 4.9 Weight loss—control sample against lap joint (screw with nut) using radiant heat source

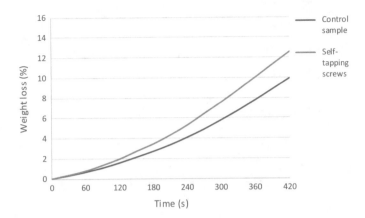

Fig. 4.10 Weight loss—control sample against lap joint (self-tapping screw) using radiant heat source

4.1.2   Flame Source

The measurements for this heat source are shown in Figs. 4.11, 4.12, 4.13, 4.14, 4.15, 4.16, 4.17, 4.18, 4.19, and 4.20. Figure 4.11 shows weight loss curves for all types of joints using flame heat source. As we can see in Fig. 4.11, the selected type of joint has an impact on the weight loss curve. The curves for radiant heat source are more convex, whereas they are rather concave for flame heat source. The weight loss values are also lower than in the case of radiant heat source. It is caused by the nature of the source. The majority of the sample's surface is exposed to radiant heat source (causing greater weight loss), on the other hand, the area around the joint exposed to flame source is more point-source (lower weight loss). That's why we did not

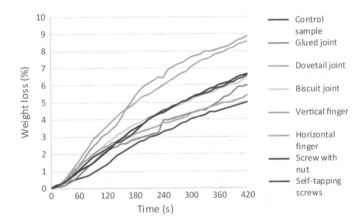

Fig. 4.11   Weight loss course of the given joints using flame source

compare these sources/the results obtained for the individual sources. If comparing the effect of the joint on weight loss under flame source, we can come up with similar conclusions as for radiant heat source.

Both figures, Fig. 4.12 (peak weight loss values) and Fig. 4.13 (the differences in peak weight loss values for each joint after taking away the peak weight loss value of the control jointless sample) show that the worst results have been achieved for dovetail and vertical finger joint samples. The differences in the case of these two joints are even more significant when using flame source. This is caused by the fact that the joint is directly exposed to fire and heat transfers through the joint (the joint's length was discussed in the previous section).

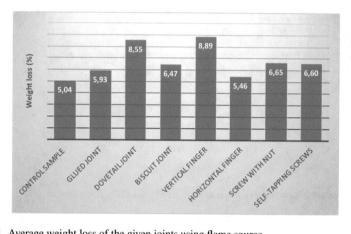

Fig. 4.12 Average weight loss of the given joints using flame source

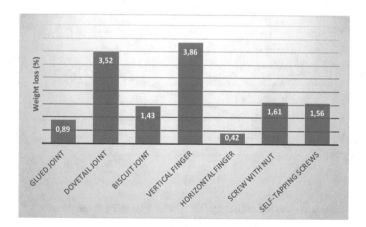

Fig. 4.13 Average differences in weight loss—a control sample against the given joint using flame source

Figures 4.14, 4.15, 4.16, 4.17, 4.18, 4.19, and 4.20 show two curves only—the curve of the control measurement for a jointless sample and for the given type of joint. These curves make the effect of the joint on the weight loss curves even more obvious. The curves for glued joint and horizontal finger joint almost follow the curve for a control (jointless) sample. As for the joints with inset elements (self-tapping screw, a screw with nut, a biscuit), the curves are very similar and in the case of dovetail joint and vertical finger joint, the differences are extreme.

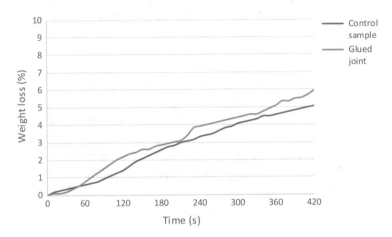

Fig. 4.14   Weight loss—control sample against glued joint using flame source

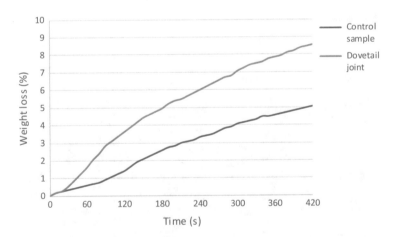

Fig. 4.15   Weight loss—control sample against dovetail joint using flame source

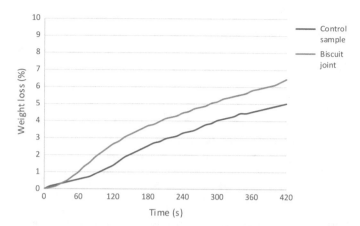

Fig. 4.16   Weight loss—control sample against biscuit joint using flame source

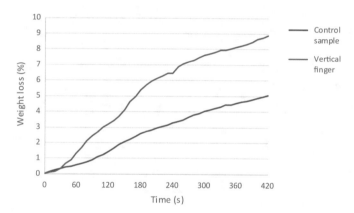

Fig. 4.17   Weight loss—control sample against vertical finger joint using flame source

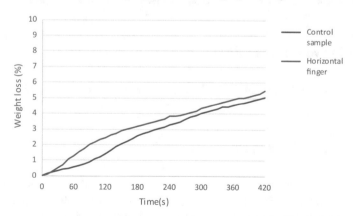

Fig. 4.18   Weight loss—control sample against horizontal finger joint using flame source

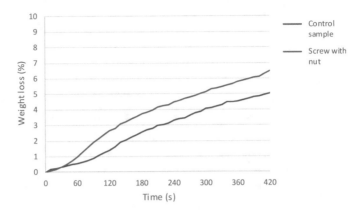

Fig. 4.19 Weight loss—control sample against lap joint (screw with a nut) using flame source

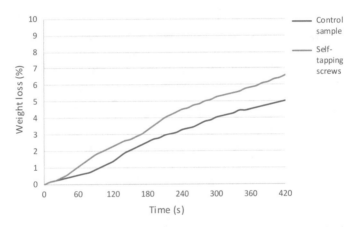

Fig. 4.20 Weight loss—control sample against lap joint with a self-tapping screw using flame source

4.2 Relative Burning Rate

Relative burning rate was calculated by deriving weight loss (see Sect. 3.5.2).

4.2.1 Radiant Heat Source

The results for radiant heat source are shown in Figs. 4.21, 4.22, 4.23, 4.24, 4.25, 4.26, 4.27, 4.28, 4.29, and 4.30. Figure 4.21 shows the curves of relative burning rate for all types of joints, Fig. 4.22 shows the peak values of the relative burning rate for all types of joints under radiant heat source and Fig. 4.23 shows the differences between

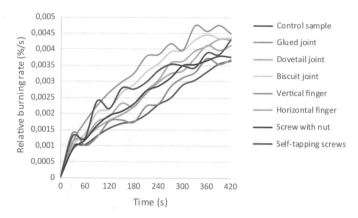

Fig. 4.21 The relative burning rate course of the given joints using radiant heat source

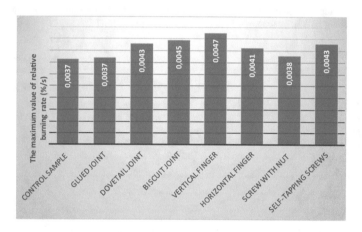

Fig. 4.22 Average peak values of the relative burning rate of the given joints

the peak burning rate value for the control (jointless) sample and the peak values of relative burning rate for each type of joint using radiant heat source. Figure 4.23 clearly shows that the biggest difference 0.0011 (%/s) is between a control sample and vertical finger joint.

Figures 4.14, 4.15, 4.16, 4.17, 4.18, 4.19, 4.20, 4.21, 4.22, 4.23, 4.24, 4.25, 4.26, 4.27, 4.28, 4.29, and 4.30 compare the relative burning rate course between a control (jointless) sample and the selected joints.

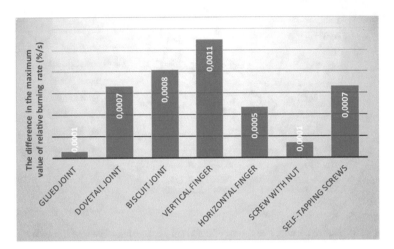

Fig. 4.23 Average values of the differences in peak relative burning rate between a control sample and a joint using radiant heat source

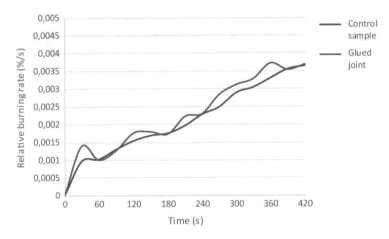

Fig. 4.24 Relative burning rate—control sample against glued joint using radiant heat source

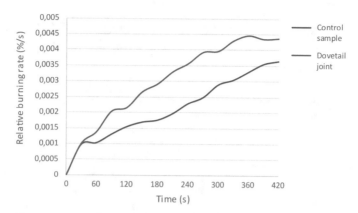

Fig. 4.25 Relative burning rate—control sample against dovetail joint using radiant heat source

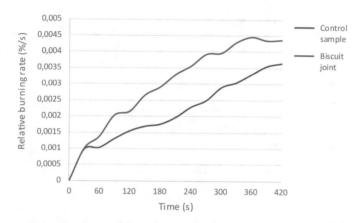

Fig. 4.26 Relative burning rate—control samples against biscuit joint using radiant heat source

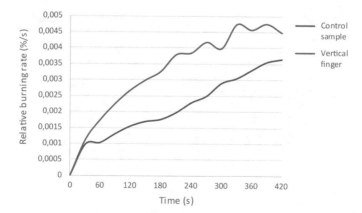

Fig. 4.27 Relative burning rate—control sample against vertical finger joint using radiant heat source

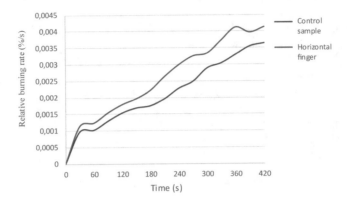

Fig. 4.28 Relative burning rate—control sample against horizontal finger joint using radiant heat source

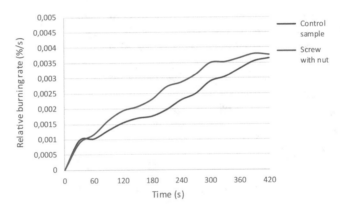

Fig. 4.29 Relative burning rate—control sample against lap joint (screw with a nut) joint using radiant heat source

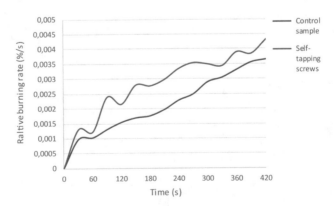

Fig. 4.30 Relative burning rate—control sample against lap joint (with a self-tapping screw) using radiant heat source

4.2.2 Flame Source

The results for flame source—relative burning rate used as the evaluation criterion—are shown in Figs. 4.31, 4.32, 4.33, 4.34, 4.35, 4.36, 4.37, 4.38, 4.39, and 4.40. Figure 4.31 shows the relative burning rate curve for all types of joints, Fig. 4.32—the relative burning rate and its peak values and Fig. 4.33 shows the differences in the peak value of the relative burning rate for a control (jointless) sample and each type of joint. If we look at Fig. 4.33 in detail, we can see that the worst results have been achieved for dovetail joint (0.0019%/s) and vertical finger joint (0.0026%/s). Glued joint and lap joint with a self-tapping screw came off best—the difference in the peak values of relative burning rate between them was null. In the case of

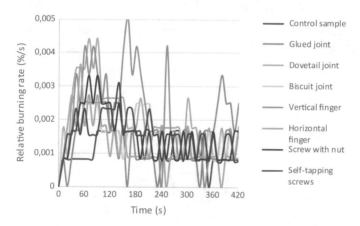

Fig. 4.31 Relative burning rate course of the given joints using flame heat source

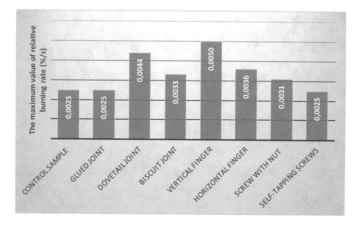

Fig. 4.32 Average values of the relative burning rate of the given joints using flame source

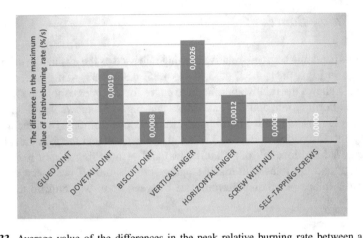

Fig. 4.33 Average value of the differences in the peak relative burning rate between a control sample and a joint using flame source

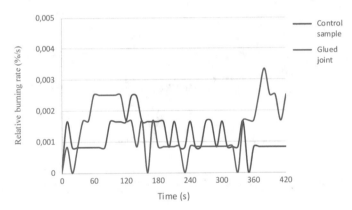

Fig. 4.34 Relative burning rate—control sample against glued joint using flame source

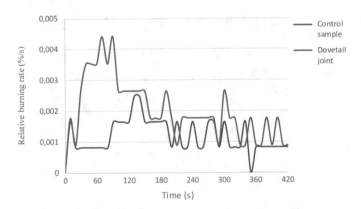

Fig. 4.35 Relative burning rate—control sample against dovetail joint using flame source

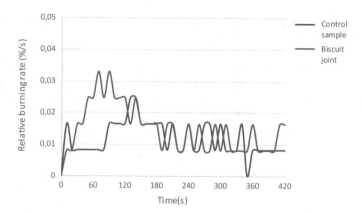

Fig. 4.36 Relative burning rate—control sample against biscuit joint using flame source

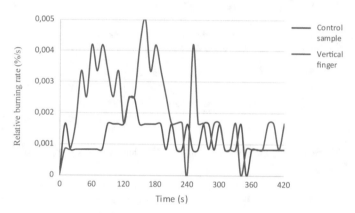

Fig. 4.37 Relative burning rate—control sample against vertical finger joint using flame source

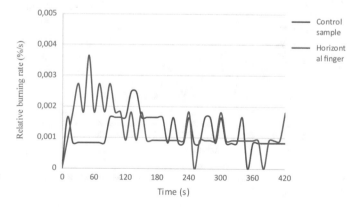

Fig. 4.38 Relative burning rate—control sample against horizontal finger joint using flame source

lap joint with self-tapping screws, the position of the tip of the flame (between the self-tapping screws) could have played an important role.

Figures 4.34, 4.35, 4.36, 4.37, 4.38, 4.39, and 4.40 compare the relative burning rate between a control (jointless) sample and the selected types of joints. As we can see from these figures, the relative burning rate course is not as homogeneous as in the case of radiant heat source. Besides the heat itself, flame source also inflicts some mechanical action—the flame "blows away" some ash particles as well as the charred layer what is reflected in the measurements.

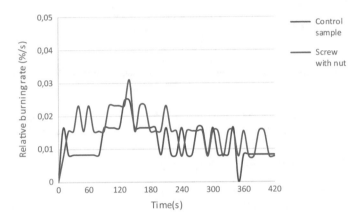

Fig. 4.39 Relative burning rate—control sample against lap joint (screw with a nut) joint using flame source

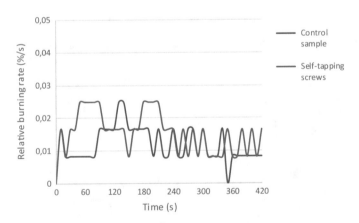

Fig. 4.40 Relative burning rate—control sample against lap joint with a self-tapping screw using flame source

4.3 Ratio P

Ratio P is a value calculated using the relative burning rate (see Sect. 3.5.3). As stated in this chapter, both the maximum burning rate value and the time when the value was recorded are important. A table of the average input and the output values is provided and the differences in the values for each joint type (without control samples) are depicted in a form of a graph. Just like with the other evaluation criteria, the two heat sources—flame and radiant heat source—are not compared with each other (only briefly commented on).

4.3.1 Radiant Heat Source

The P ratio values for the radiant heat source are shown in Table 4.1 and Fig. 4.41. The values range from 8.69 for the control sample up to 13.19 for the vertical finger joint. The maximum burning rate values for the radiant heat source were achieved at the end of the experiment which had an impact on the final P ratio value. If we look at each joint separately, vertical finger joint achieved the worst results for this evaluation criterion. In case of lap joint with a self-tapping screw, a relatively high value has been recorded. All the other joints have the same ranking as in the other evaluation criteria.

Table 4.1 The data on the peak burning rate, the time necessary to achieve it and the P ratio for the given joints using radiant source of heat

Joint	Max. burning rate (%/s)	Time max. burning rate (s)	Ratio P $((\%/s^2).10^6)$
No joint—control sample	0.00457	600	7.62
Glued joint	0.00406	510	7.96
Dovetail joint	0.00467	510	9.16
Biscuit joint	0.00481	450	10.69
Vertical finger	0.00475	390	12.18
Horizontal finger	0.00464	510	9.10
Screw with nut	0.00411	510	8.06
Self-tapping screws	0.00483	600	8.05

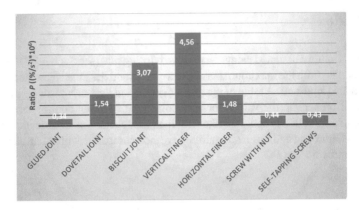

Fig. 4.41 The differences in the average peak burning rate between the given joint and a control sample using radiant heat source

4.3.2 Flame Heat Source

P ratio values for the flame source are set out in Table 4.2 and Fig. 4.42. *P* ratio values range from 19.08 for the control sample up to 69.83 for the vertical finger joint. As it is obvious from the results, the values (using the same calculations) are significantly higher due to the time when the maximum relative burning rate was achieved, which is significantly lower than for radiant heat source. Maximum burning rate for flame source is growing the moment the sample is exposed to the flame of the burner. The surface of the sample catches fire, a charred layer is created on the surface while the sample is still burning, but it will not reach the maximum relative burning rate value, which was recorded earlier, at the initial stage of the experiment. If we compare the joints individually, the worst result for this evaluation criterion was recorded for vertical finger joint samples.

The information value of this criterion can be even more obvious in case of the lap joint with a self-tapping screw—when using flame heat source, the maximum relative burning rate value for this joint is equivalent to the one for the control sample—0.0025%/s (see Table 4.2 as well as Fig. 4.42). If we take into account the time factor, the *P* ratio value changes drastically—a relatively high value of this evaluation criterion—the third highest one—has been measured. All the other joints have the same ranking as in the other evaluation criteria.

The best results were achieved for horizontal finger joint and glued joint while using this evaluation criterion. Lap joint with screws and a nut has reached the best value within this evaluation criterion 22.07 ($(\%/s^2).10^6$). We just need to draw attention to the "flaws" of the experiment, since the metal fasteners were not directly exposed to the flame. The flame was directed between these two fasteners and therefore the joint could not be objectively assessed as the best one. If the metal fastener had been exposed to a direct flame, the result wouldn't have been this positive.

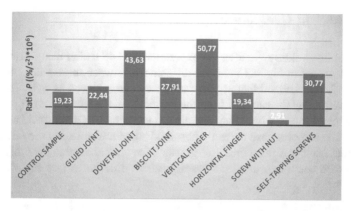

Fig. 4.42 The differences in the average peak burning rates between the given joint

Table 4.2 The data on the peak burning rate, the time necessary to achieve it and the P ratio for the given joints using flame heat source

Joint	Max. burning rate (%/s)	Time max. burning rate (s)	Ratio P ((%/s^2).10^6)
No joint—control sample	0.0025	130	19.23
Glued joint	0.0013	60	22.44
Dovetail joint	0.0031	70	43.63
Biscuit joint	0.0020	70	27.91
Vertical finger	0.0030	60	50.77
Horizontal finger	0.0014	70	19.34
Screw with nut	0.0004	140	2.91
Self-tapping screws	0.0015	50	30.77

4.4 Weight Loss Difference 0–48

Weight loss difference 0–48 is a value calculated using the average values for each joint i.e. the weight loss value right after the experiment and 48 h after the experiment (see Sect. 3.5.4) to observe the occurrence of any follow-up burning or flameless burning—smoldering of spruce wood, depending on the type of joint, after finishing a continuous measurement of weight loss. Here are the average values and the differences in values for individual joints including the jointless control sample in a form of a graph. Just as with the other evaluation criteria, the two heat sources—flame and radiant heat source—are not compared (only briefly commented on).

4.4.1   Radiant Heat Source

The weight loss values right after the experiment and 48 h after using a radiant heat source are shown in Fig. 4.43 and the differences between those values are shown in Fig. 4.44. We assumed that negative values can be achieved as well. This means that after removing the samples from the heat source, the samples were put into an air-conditioned room, where they can absorb some moisture and their weight can increase. The samples have certainly absorbed some moisture, but the effect of the radiant heat itself was in all the cases greater even after absorbing the moisture. We have detected small but, in all the cases, positive differences (see Fig. 4.44). The

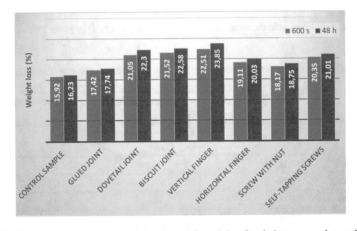

Fig. 4.43 The average weight loss of the given joints right after being exposed to radiant heat source and 48 h after the exposure

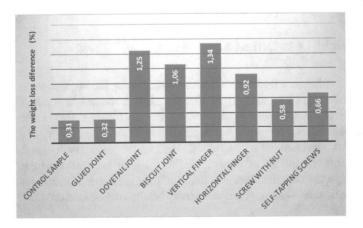

Fig. 4.44 The differences in the average weight loss values of the given joints right after being exposed to a radiant heat source and 48 h after the exposure

differences ranged from 0.31 up to 1.34%. The worst results were achieved for the horizontal finger joint and the dovetail joint. On the other hand, the best results were achieved for the control sample and the glued joint. Samples with metal fasteners achieved relatively good results as well. This could be caused by the weight of the fasteners which didn't change over time or during the experiment.

4.4.2 Flame Heat Source

Higher values (see Figs. 4.45 and 4.46 have been detected when using flame heat source. The differences between the weight losses (0 and 48) ranged from 1.04% for

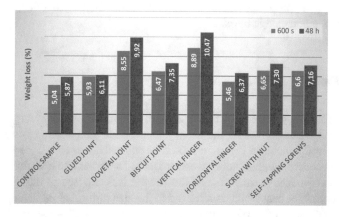

Fig. 4.45 The average weight loss of the given joints right after being exposed to flame heat source and 48 h after the exposure

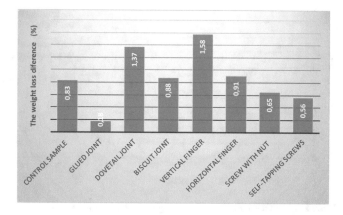

Fig. 4.46 The differences in average weight loss values for the given joints right after being exposed to a flame heat source and 48 h after the exposure

the control (jointless) sample up to 4.89% for vertical finger joint. For this evaluation criterion, the flame easily penetrated the joint itself directly causing that a thicker layer of the wood was affected by the flame and the highest weight loss values could be observed.

4.5   Charred Layer

The size of charred layer was calculated as referred to in Sect. 3.5.5. All the samples have been cut in the middle and the charred area and pyrolytic layer were measured. Based on these measurements, the size of individual layers was calculated. The values for all the samples were averaged and the result—the diameter—is incorporated in the charts that are presented in this chapter. Subsequently, the size of each layer was evaluated on a percentage basis.

4.5.1   Radiant Heat Source

Figures 4.47, 4.48, 4.49, 4.50, 4.51, 4.52, 4.53, and 4.54 show the average values of pyrolytic and charred layer for the given joints. Figure 4.55 shows pyrolytic layers for all the given joints, and Fig. 4.56 shows the charred layers for all the joints using radiant heat source. Higher values were recorded for radiant heat source compared to flame heat source. Greater area of the sample was exposed to the thermal load when using radiant heat which is reflected in the size of its charred layer.

The differences in the values between the joints are small—due to the material used rather than the joint itself—therefore we will not evaluate and comment on it

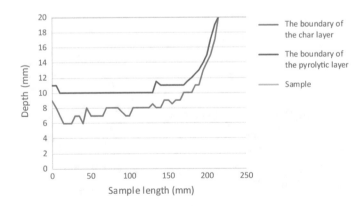

Fig. 4.47 Charred layer of a jointless samples using radiant heat source

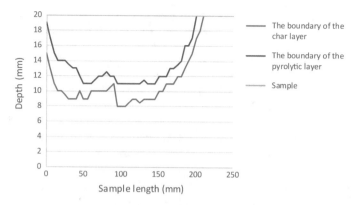

Fig. 4.48 Charred layer of a glued joint sample using radiant heat source

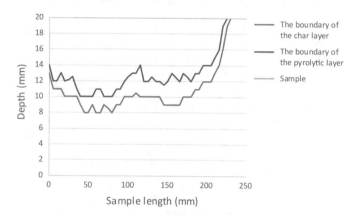

Fig. 4.49 Charred layer of dovetail joint samples under radiant heat source

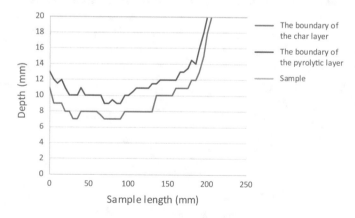

Fig. 4.50 Charred layer of biscuit joint samples using radiant heat source

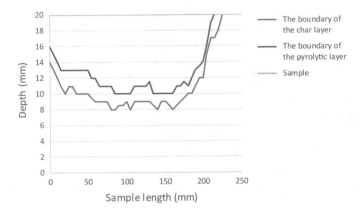

Fig. 4.51   Charred layer of vertical finger joint samples using radiant heat source

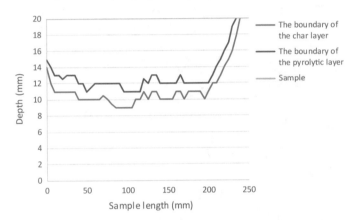

Fig. 4.52   Charred layer of horizontal finger joint samples using radiant heat source

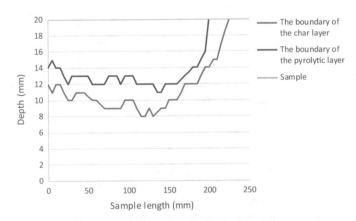

Fig. 4.53   Charred layer of lap joint with a screw and a nut using radiant heat source

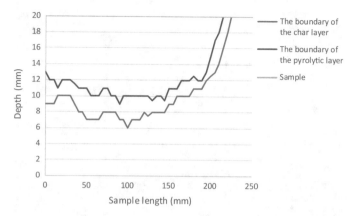

Fig. 4.54 Charred layer of lap joint samples with a self-tapping screw using radiant heat source

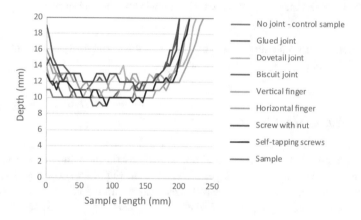

Fig. 4.55 The edges of the pyrolytic layer for the given joints using radiant heat source

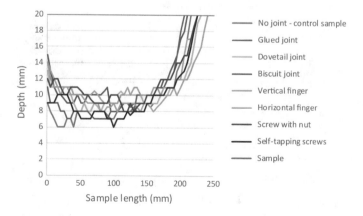

Fig. 4.56 The edges of the charred layer for the given joints using radiant heat source

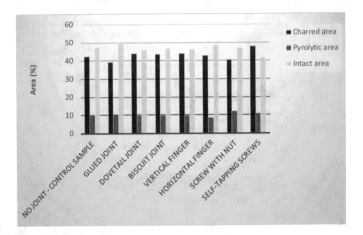

Fig. 4.57 The layers and their percentage for the given joints using radiant heat source

any further. Figure 4.57 shows the percentage of charred, pyrolytic and intact layer for the given joints using radiant heat source.

4.5.2   Flame Heat Source

Figures 4.58, 4.59, 4.60, 4.61, 4.62, 4.63, 4.64, and 4.65 show the average values of the pyrolytic layer and charred layer for each joint, Fig. 4.66 shows pyrolytic area for all the joints, and Fig. 4.67 shows charred area for all the joints under flame heat source. Lower values were recorded when a flame heat source was used. As Fig. 3.13 shows, the flame of the burner operates in the middle of the sample (where the joint

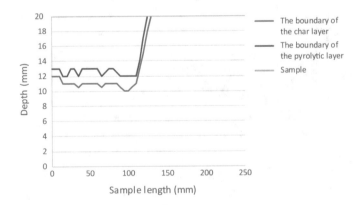

Fig. 4.58 Charred layer for jointless samples using flame source

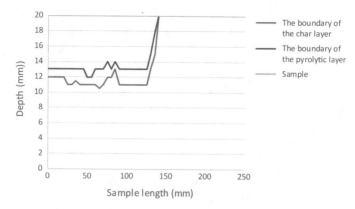

Fig. 4.59 Charred layer for glued joint samples under flame source

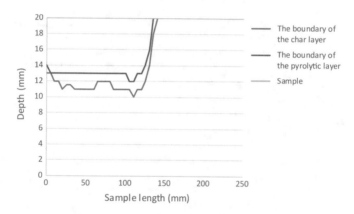

Fig. 4.60 Charred layer for dovetail joint samples under flame source

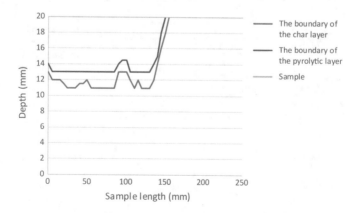

Fig. 4.61 Charred layer for biscuit joint samples under flame source

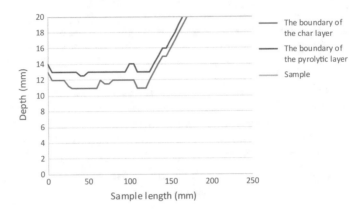

Fig. 4.62   Charred layer for vertical finger joint samples under flame source

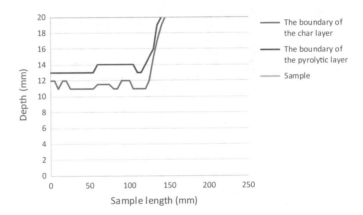

Fig. 4.63   Charred layer for horizontal finger joint samples under flame source

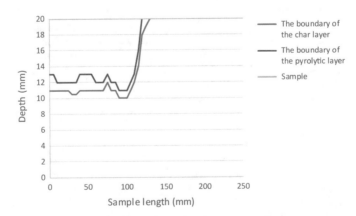

Fig. 4.64   Charred layer for lap joint samples with a screw and a nut under flame source

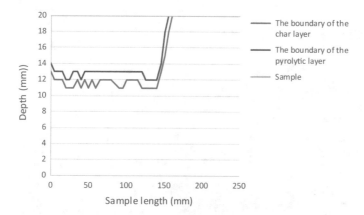

Fig. 4.65 Charred layer for lap joint samples with a self-tapping screw under flame source

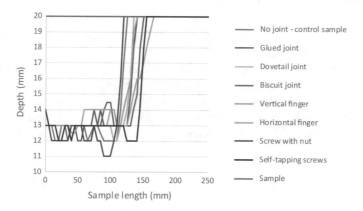

Fig. 4.66 The edge of the pyrolytic layer for the given joints under flame source

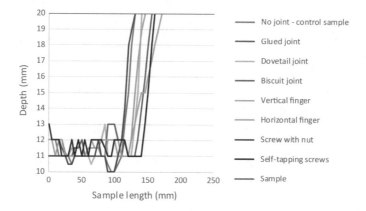

Fig. 4.67 The edges of the charred layer for the given joints using flame source

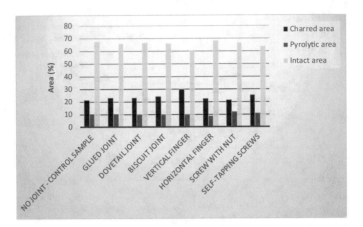

Fig. 4.68 The layers and their percentage for the given joint using flame source

is) and only the upper part of the sample is exposed to the flame causing that pyrolytic and charred layers are smaller (see Fig. 4.58).

We would like to draw attention to Figs. 4.51 and 4.52 which show the charts for horizontal and vertical finger joint under this type of thermal load. If we compare the two charts (including Chart Fig. 4.51 with the other graphs) we can see that the charring course can be detected even past the middle of the sample where the flame operated. The selected evaluation criterion counts against this joint type (Fig. 4.68).

Chapter 5
Evaluation and Discussion

Abstract This chapter compares the results obtained from the experiments with those of other authors.

Keywords Spruce wood · Wood density · Thermal modification of wood · Wood certification · Changes in the wood properties · Wood joints

The implementation of this (relatively difficult) experiment is justifiable. As stated in Chap. 1, fire façades are quite common even in cases when buildings are faced with non-combustible materials. The main problem is the position of sheathing and the facing, which is identical with the rapid spread of the fire. If we use flammable materials for sheathing or the thermal-insulation system, we run the risk of a potential fire which could spread quickly and have serious consequences due to the combination of the material and its position within the structure. This issue is treated by Iringova in her works [1–4]. However, in the case of ecological buildings or zero-energy houses, we mainly opt for natural materials which can catch fire and burn. Elimination of the material in construction industry is not a way to go. On the contrary, we have to search for a way of modifying its properties and verifying the possibility for their application in compliance with the safety rules [5, 6]. In addition to these modifications, it is important to test the modifications, to assess their characteristics which have to be enshrined in the legal and technical regulations. By means of certification, wood and wood-based materials can be used in practice [7].

Wood-based sheathing and cladding materials can be not only used in wooden structures but also in structures incorporating other construction elements. Wood, thanks to its natural and unique character, provides architects with numerous opportunities to make use of their artistic skills. In order for the buildings to be safe, it is necessary to apply certified materials in accordance with national or international regulations [8, 9]. Before the material is subjected to certification, it undergoes many scientific experiments. Wood has been subjected to such experiments many times. Various native tree species [10–13] as well as tropical tree species [14, 15]. A great deal of attention has been paid to spruce as an important Central European tree species and the most important tree species in Slovakia in economic terms [16–19]. The spruce is studied in detail, not only in its natural form, but fire-retardant modifications

of different products are observed [20–22] including the effect of spruce wood modification (thermal treatment—Thermowood) on its fire-technical properties [23–30].

Two factors have been subject to studies in the experiment—physical characteristics of spruce wood and their impact on its combustion and ignition. The test
methods that would be appropriate for the experiment evaluation were also studied.
After getting to know the basic information on fire technical properties of spruce
wood [31–39], we could proceed to the experiment. Out of the test methods available, we chose different ones [31–33, 40–46] monitoring the conditions of thermal
load, the type of source and the exposure time, the sample's parameters, the evaluation criteria and the results achieved. On the basis of these studies, the test conditions
were modified [47, 48] and experiment parameters were created so that the effect of
joints can be properly evaluated using the given evaluation criteria. We have modified the conditions as follows: an emerging fire, a source of lower intensity and a
relatively short time of exposure.

First, radiant heat source was chosen and scenario A was observed (see Fig. 1.5,
Chap. 1). Later, flame heat source was used. Their exposure time was the same and
relatively short—420 s. The time frame should suffice to be reflected in the evaluation
criteria, and, at the same time, it will not degrade the samples completely, they won't
burn away completely, and there will be no need to extinguish them. As the results
have shown, the time exposure of the samples to a heat source as well as its intensity
proved themselves to be accurate. The next condition of the experiment was an "open
space". It was a laboratory which wasn't surrounded with any auxiliary equipment in
order to create the most accurate simulation of potential fire conditions in the exterior.

Great attention was paid to the choice of the samples. Besides dimensions, their
density played an important role as well [49, 50]. On the basis of some works as
well as previous studies concerning spruce wood burning, we have paid a great
deal of attention to density. Air-conditioning did not pose any problem, the problem
lied in choosing the density level. A large number of samples had to be sorted out
according to their quality, defects but mainly according to their density. Density
interval fluctuated ± 5 kg/m^3, which was a very low value (see Chap. 3, Table 3.1).
Having the samples sorted out, the samples incorporating the given joints were made.
Attention was also paid to the dimension of samples. They had to be "large" enough
so that the joint could manifest itself. Their dimensions were interconnected with the
density, as it would be difficult and economically very costly to sort large samples
out (by density). The selected sample sizes have proven to be adequate not only for
density classification, but also for the experiment itself and when finding out the
effect of a joint according to the selected evaluation criteria.

A great deal of attention was paid to the selection of the evaluation criteria. Weight
loss is one of the oldest evaluation criteria when assessing the fire characteristics of a
material. It is applied as an evaluation criterion in many other sophisticated laboratory
facilities and procedures evaluating materials in relation to their ignition, burning,
flameless burning, in other words to fire—technical properties of the tested materials.
Weight loss has been recorded continuously [51, 52].

Based on the derivation of weight loss, the burning rate and ratio P was calculated.

As we have stated before, not only the maximum relative burning rate value is important, it is also the time, when this peak value was recorded. Ratio P represents the percentage of these values.

The difference in weight loss marked as 0–48 represented another evaluation criterion. We wanted to find out the behavior of samples in an air-conditioned room after the test, in particular its change in weight loss. Will the samples burn after removing it from the heat source? Will flame or flameless burning occur? Will it have an impact on weight loss? Will the given joint play any role in that? As it turned out, this evaluation criterion was chosen correctly.

The last evaluation criterion was the size of charred layer on the samples after cutting them in the middle. Charred layers and charring are in the center of attention of several authors [53–55], and this factor is taken into account in Eurocode 5 as well [56].

Brief comments on each result were given in the previous chapter. If we generalize the results of all measurements taking into consideration all evaluation criteria, we can observe some differences in measured values. Sometimes minor ones (we didn't expect any major ones given the conditions of the experiment), and in some cases, the differences have been greater. It is essential that the jointless sample always reached the best results. The values of other evaluation criteria for other joints can be derived from that. The dovetail joint and the vertical finger joints achieved the worst results. As we summed it up in Sect. 4.1.1, these joints allow heat to be transferred directly to the other side of the sample. The dovetail itself is 160 mm long and the indentation is up to 360 mm long in the case of vertical finger joint. This fact certainly affects the weight loss of the given joints, even though the sample is fully glued. In terms of the thickness, there is no direct obstacle to heat transfer in these joints. The entire indentation area is exposed to heat, its intensity gradually goes up for both sources, which manifested itself in the evaluation criteria.

The other joints do not show between themselves any significant differences within the evaluation criteria. This means that this issue has to be addressed. It seems to be appropriate to carry out and improve a similar experiment by measuring the temperature inside the joint or to carry out a large-scale test. In addition to crosswise joints, attention should be paid to longitudinal joints as well. Large-scale tests based on fire model tests [57–59], or the standardized tests [42, 60–64] will provide additional and necessary information on the behavior of wood connected to using woodworking joints and combined woodworking joints (with a different fastener) in the case of a fire.

References

1. Iringova A (2018) Design of envelopes for timber buildings in terms of sustainable development in the low-energy construction. In: IOP conference series: Mater Sci Eng 415, art no 012010: 1–8. https://doi.org/10.1088/1757-899x/415/1/012010, Accessed on 22 Nov 2005. ISSN: 1757-8981

2. Iringova A (2017) Impact of fire protection on the design of energy efficient and eco-friendly building envelopes in timber structures. In: Fire protection, safety and security 2017 [CD ROM]: conference proceedings. Zvolen: Technical university in Zvolen, 2017, p. 58–64. ISBN: 978-80-228-2957-1

3. Iringova A (2017) Lightweight building envelopes in prefabricated buildings in terms of fire resistance. In: MATEC Web of conferences, 2017, vol 117, art no 00062. ISSN: 2261 236X

4. Iringova A (2017) Revitalisation of external walls in listed buildings in the context of fire protection. In: Procedia Engineering,vol 195, pp 163–170. ISSN: 1877-7058, 2017

5. Martinka J, et al (2013) New trends in fire risk assessment of lignocellulosic materials using a conical calorimeter. [Nové trendy posudzovania požiarneho rizika lignocelulózových materialov kónickým kalorimetrom]. In: Advances in Fire & Safety Engineering, [CD ROM]: conference proceedings. Zilina: University of Zilina, 2013, pp 57–64. ISBN: 978-80-88829-80-5

6. Mitterova I, et al (2014) The comparison of flame retardants efficiency applied on spruce and chipboard samples exposed to the radiant and flame sources. In: Advances in Fire, Safety and Security Research 2014. [Bratislava]: Fire Research Institute of the Ministry of Interior SR, 2014, pp 31–40. ISBN: 978-80-89051-16-8

7. Cholin JM (2008) Wood and wood-based products. In: Fire protection handbook. Second edition. Quincy: National fire protection association, s 6.61–6.74. ISBN: 978-0-87765-758-3

8. EN 13501-1: 2007+A1 (2009) Fire classification of construction products and building elements. Classification using test data from reaction to fire tests

9. Alternative Solution Fire Compliance Façades (2002a) Technical Design Guide issued by Forest and Wood Products Australia. Prepared by: Exova Warrington fire Aus. Pty Ltd Suite 2002a, Level 20, 44 Market Street Sydney 2000 Australia First published: June 2013, ISBN: 978-1-921763-68-7

10. Osvald A (2017) Weight loss as an assessment criterion for fire related properties. In. Advances in fire and safety engineering 2017, Trnava: Alumni Press Trnava, 2017, pp 10–15. ISBN: 978-80-8096-245-6

11. Martinka J, et al (2017) Ignition parameters of poplar wood. Acta Facultatis Xylologiae 59(1): 85–95. ISSN: 1336-3824

12. Metsä-Kortelainen S, Viitanen H (2009) Decay resistance of sapwood and heartwood of untreated and thermally modified Scots pine and Norway spruce compared with some other wood species. Wood Mater. Sci. Eng. 4(3–4): 105–114. https://doi.org/10.1080/174802709033 26140. ISSN: (printed): 1748-0272

13. Sotomayor CJM, Galllegos GL (2018) Reacción al fuego de madera sólida y laminada de Enterolobium cyclocarpum, Tabebuia rosea y Juniperus pyriformis. Estudio comparativo. In: Investigación e IIM Ingeniería de la Madera 4(1): 39–78. ISSN: 2395-9320

14. Makovicka Osvaldova L, et al (2018) Effect of thermal treatment on selected fire safety features of tropical wood. In: Communications: scientific letters of the University of Zilina 20(2): 3–7. [print] ISSN: 1335-4205

15. Kadlicova P, et al (2018) Tropical wood facing material under fire Conditions [electronic] In: Book of proceedings international scientific conference Earth in a trap? 2018 [electronic]: Analytical methods in fire and environmental sciences. 1. Edition, Zvolen: Technical university in Zvolen, 2018, pp 87–90. ISBN: 978-80-228-3062-1

16. Chovanec D, Osvald A (1992) Spruce wood structure changes caused by flame and radiant sources. Zvolen: Technical university in Zvolen, 1992, 62 p. ISBN: 978-80-228-1034-7

17. Markova I (2000) Wood thermal degradation. Observation of changes in saccharide portion of thermally degraded spruce wood by gas chromategraphy. In Wood and fire safety: proceedings. Part 1. Zvolen: Technical university in Zvolen, 2000, pp 195–200. ISBN: 80-228-0774-5

18. Ruzinska E, et al (2015) Modified method of evaluation the effect of the protection retardant coatings applied on the surface of the wood building construction. In: Materials science forum, 2015, vol 818, pp 194–197. ISSN: 1662-9752

19. Cabalova I, et al (2013) The influence of radiant heating on chemical changes of spruce wood. In: Acta Facultatis Xylologiae 56(2): 59–66. ISSN: 1336-3824

20. Troitzsch J (2000) Fire gas toxicity and pollutants in fires-The role of flame retardants. In Flame retardants 2000, February 8–9, 2000, London, pp 177–184. Interscience Communications, London, UK. ISBN: 0 9532312 4 0

21. Mitterova I, Zachar M (2013) The comparison of flame retardants efficiency when exposed to heat. In: Modern trends in ergonomics and occupational safety. Zielona Góra: University of Zielona Góra, 2013. ISBN: 978-83-7842-086-6, s 207–226

22. Mitterova I, et al (2014) Ignitability of unprotected and retardant protected samples of spruce wood. In: Advanced materials research, 2014, vol 1001, 2014, pp 330–335. ISSN: 1022-6680

23. Cekovska H, et al (2017) The effect of thermal treatment of spruce wood on its fire performance characteristics. In: Annals of Warsaw university of life sciences-SGGW: Forestry and wood technology, no 97. ISSN: 1898-5912

24. Metsä-Kortelainen S, et al (2011) Durability of thermally modified Norway spruce and Scots pine in above ground conditions. Wood Material Science and Engineering 6(4): 163–169. https://doi.org/10.1080/17480272.2011.567338. ISSN: (printed): 1748-0272

25. Yildiz, S. et al. (2006): Mechanical and chemical behaviour of spruce wood modified by heat. Build Environ 41(12): 1762–1766. https://doi.org/10.1016/j.buildenv.2005.07.017. ISSN: 0360-1323

26. Sikora A et al (2017) Selected physical properties of thermally modified spruce wood. In: PRO LIGNO 13(4): 123–132. ISSN: 1841- 4737

27. Kacikova D, et al (2013) Effects of thermal treatment on chemical, mechanical and colour traits in Norway spruce wood. Bioresource Technology 2013, 144–669. https://doi.org/10.1016/j.biortech.2013.06.110. ISSN: 1930-2126

28. Kacik F, et al (2012) Release of terpenes from fir wood during its long-term use and in thermal treatment. Molecules 17 (8): 9990–9999. https://doi.org/10.3390/molecules17089990. ISSN: 1420-3049

29. Martinka J, et al (2013) An examination of the behaviour of thermally treated spruce wood under fire conditions. Wood Res 58 (4): 599–606. ISSN: 1336-4561

30. Barvik Š, et al (2015) Effect of temperature on the color changes of wood during thermal modification. Cell Chem Technol 49(9–10): 789–798. ISSN: 0576-9787

31. Markova I, Klement I (2003) Thermal analysis (TG, DTG and DSC) of hornbeam wood afterdrying. Wood Res 48(1–2): 53–61. ISSN: 0012-6136

32. Oremusova E, et al (2014) Evaluation of the gross and net calorific value of the selected wood species. In: Advanced Materials Research, 2014, vol 1001, pp 292–299. ISSN: 1022-6680

33. Geffert A, et al (2017) Swelling of cellulosic porous materials–mathematical description and verification. In BioResources, n 3, pp 5017–5030. ISSN: 1930-2126

34. Siklienka M, et al (2016) The influence of milling heads on the quality of created surface. In Acta Facultatis Xylologiae, Zvolen: Technical university in Zvolen, 2016, no 2, pp 81–88. ISSN: 1336-3824

35. Dzurenda L, Jandacka J (2010) Energy use of dendromass. [Energetické využitie dendromasy] Zvolen: Technical university in Zvolen, 2010, (monograph), 162 p. ISBN: 978-80-228-2082-0

36. Bekhta P, Niemz P (2003) Effect of high temperature on the changes in colour, dimensional stability and mechanical properties of spruce wood. Holzforschung 57(5): 539–546. https://doi.org/10.1515/HF.2003.080. ISSN: 0018-3830

37. Welzbacher CR, Rapp AO (2007) Durability of thermally modified timber from industrialscale processes in different use classes: Results from laboratory and field tests. Wood Mater Sci Eng 2(1): 4–14. https://doi.org/10.1080/17480270701267504. ISSN: (printed): 1748-0272

38. Martinka J, et al (2016) Investigation of the influence of spruce and oak wood heat treatment upon heat release rate and propensity for fire propagation in the flashover phase. Acta Facultatis Xylologiae Zvolen 58(1): 5–14. https://doi.org/10.17423/afx.2016.58.1.01. ISSN: 1336-3824

39. Williamson RB, Baron FM (1973) A corner fire test to simulate residential fires. J Fire Flammabil 4: 99–105. ISSN: 0022-1104

40. Tran HC, White RH (1992) Burning rate of solid wood measured in a heat release rate calorimeter. Fire and Materials, no. 16, 1992, pp 197–206. ISSN: 0308-0501

41. Nuopponen M, et al (2003) The effect of a heat treatment on the behaviour of extractives in softwood studied by FTIR spectroscopic methods. In: Wood Science and technology 37, 2003, pp 109–115. ISSN: 0043-7719

42. Hasemi Y (1979) Flashover criteria of compartment fire-theory on zero order reaction system. (BRI Research Paper No. 83). Tsukuba, Japan: Building Research Institute
43. Tran HC, White RH (1992) Burning rate of solid wood measured in a heat release rate calorimeter. Fire and Materials, no.16, 1992, pp 197–206. ISSN: 0308-0501
44. Mikkola E (2009) Ignitability of Solid Materials. In Babrasukas V, Grayson SJ (eds) Heat relase in fires. Interscience Comunicatons, 2009, pp 225–232. ISBN: 978-09556548-4-8
45. Todaro L, et al (2015) Thermal treatment modifies the calorific value and ash content in some wood species. In: Fuel 140(1): 1–3. ISSN: 0016-2361
46. Kučera P, et al (2009) Fire engineering: fire dynamics. [Požarní inženýrství: dynamika požáru]. Ostrava: SPBI, 2009, 152 p. ISBN: 978-80-7385-074-6
47. Makovicka Osvaldova L, et al (2014) "New" method of wood fire retardants evaluation. ["Nové" metódy hodnotenia retardérov]. In: Advances in fire, safety and security research 2014: scientific book. Bratislava: Fire Research Institute of the Ministry of interior SR, 2014, pp 24–30. ISBN: 978-80-89051-16-8
48. Osvald A, Makovicka Osvaldova L (2016) New methods in the evaluation of flammability properties. [Nové metódy hodnotenia požiarno-technických vlastností]. In: Production management and engineering sciences. Leiden: CRC Press/Balkema, 2016, pp 503–507. ISBN: 978-1-138-02856-2
49. Makovicka Osvaldova L, et al (2016) The influence of density of test specimens on the quality assessment of retarding effects of fire retardants. Wood Res 61(1): 35–42. ISSN: 1336-4561
50. Osvald A, Balog K (2017) Wood burning. [Horenie dreva]. 1st edition-Zvolen: Technical University in Zvolen, 2017, 106 p. ISBN: 978-80-228-2953-3
51. Kadlicova P, et al (2017) Monitoring of weight loss of fibreboard during influence of flame. In: Procedia Engineering, vol 192 (2017), online, pp 393–398, http://linkinghub.elsevier.com/retrieve/pii/S1877705817326140. ISSN: 1877-7058, print ISBN: 978-80-554-1328-0
52. Gaspercova S, Makovicka Osvaldova L (2017) Influence of surface treatment of wood to the flame length and weight loss under load single-flame source. In: Structural and mechanical engineering for security and prevention. [S.l.]: Trans Tech Publications, 2017, pp 353–359. ISBN: 978-3-0357-1236-0 (Key engineering materials, vol 755, ISSN: 1013-9826)
53. Kucerova V, et al (2016) The effect of chemical changes during heat treatment on the color and mechanical properties of fir wood. In: BioResources 11(4): 9079–9084. ISSN: 1930-2126
54. Harada T (1996) Effects of density on charring and mass loss rate in wood specimens. In: Wood and Fire Safety, 3rd Int Scientific Conference, 1996, Slovak Republic, pp 149–156. ISBN: 80-228-0493-2
55. White RH (1988) Charring rates of different wood species. PhD dissertation, University of Wisconsin, Madison, WI, 1988, (unpublished)
56. EN 1995-1-2 (2004) Eurocode 5. Design of timber structures. General. Structural fire design
57. Buchanan A (2001) Structural design for fire safety. West Sussex: Wiley. ISBN: 0471-89060-X
58. Wald F, Kallerova P (2009) Fire test on experimental building in Mokrsko-test report. (Požarní zkouška na experimentalním objektu v Mokrsku-zpráva ze skoušky). (unpublished)
59. Thomas PH (1979) Some problem aspects of fully developed room fires. In: Fire safety and Standards. Philadelphia: American society for testing and materials, 1977, pp 112–306
60. Dillon SE.: Analysis of the ISO 9705 Room/Corner Test: Simulations, Correlations and Heat Flux Measurements. https://tsapps.nist.gov/publication/get_pdf.cfm?pub_id=916661
61. Janssen, ML Simple Model of the ISO 9705 Ignition Source, in Annual Conference on Fire Research, pp 84–88, October 28–31, 1996, Gaithersburg, Maryland, NISTIR 5904, National Institute of Standards and Technology, Gaithersburg, Maryland, 1996
62. Lee BT (1985) Standard room fire test development at the National Bureau of Standards. In: Fire safety: Science and Engineering. Harmathy TZ (ed) (ASTM STP 882). Philadelphia, PA: American society for testing and materials, 1985, pp 29–44
63. Sundström B (1986) Full scale fire testing of surface materials. (SP-RAPP 1986:45). Borås: swedish national testing Institute
64. Hasemi Y (1979) Flashover criteria of compartment fire-theory on zero order reaction system (BRI Research Paper No. 83). Tsukuba, Japan: Building Research Institute, 1979

Conclusion

Each building is a materialization of some ideas, or artistic feeling of architects into reality so it can serve its user as long as possible. Material selection depends on the purpose of the construction. Wood is one of the first and therefore basic building materials. When people stopped living in a cave built by nature, they started to build their homes—structures, which were made of wood. Wood is a building material, which is firm and easy to dress. If we are to be objective, every material has its pros and cons. The negative characteristics of wood are mainly its low biological resistance and its ability to catch fire and burn. These negative characteristics, however, are also not about the wood only. It is always the relationship of wood and people. Wood does not biotically degrade if people create a microclimate, which is appropriate for both their life and the life of wood. If they fail to treat it properly, it does not create a positive microclimate, and wood absorbs moisture and degrade biotically. The same goes for its response to fire and its ability to catch fire and burn. There are, however, limitations enshrined in fire regulations—security measures when using flammable materials.

These "limitations"—fire-fighting legislation—are subject to constant improvement. This improvement is heading in two directions: fire investigation, analysis of fire and its spread, and scientific research. The latter is the method we used when addressing the problem that is described in the monograph. At the beginning we have pointed out that the solution to the problem of joints and their effect on a fire seemed to be of peripheral importance in general. More test methods—the old as well as the modern ones—using radiant and a flame heat source have been taken into account, but none of them could or were allowed to test samples containing a joint of two materials—the same or different ones. Therefore, we chose this issue to be the objective of the research, to revise the impact of joint of two elements used within a façade on the formation and development of fire. Even though there is no real fire in the true sense of the word happening in our experiment, it is the heating, which may cause a fire.

To meet this objective, besides the joint itself, it was necessary to pay attention to the choice of the material. As the samples are subjected to the experiment for only a short period of time, the thermal load for both sources was only 7 min. Great

L. Makovicka Osvaldova, *Wooden Façades and Fire Safety*,
SpringerBriefs in Fire, https://doi.org/10.1007/978-3-030-48883-3

emphasis was put on the quality of the wood, the samples were made from. Wood had to be defect less and of low density tolerance so that wood quality does not have an effect on the evaluation criteria. The conditions of the experiment were chosen so as to simulate the conditions of an actual fire.

An open space in the laboratory has been chosen for the experiment, not an enclosed chamber or furnace.

The main criterion (weight loss) has been examined from several perspectives, along with the relative burning rate, its peak value, the time to achieve the peak value and the ratio P. All these measurements were obtained continuously when the sample was exposed to thermal load. Weight loss was recorded in 15 second intervals. Based on this value, other parameters were calculated. Two evaluation criteria were monitored discontinuously, i.e. after the sample has cooled down, weight loss 0-48 and the size of charred layer.

Eight types of joints were subjected to the experiment. The control sample without a joint, lap joint—glued joint, dovetail joint, biscuit joint, horizontal joint, vertical joint, lap joint with a screw and a nut and joint with self-tapping screws. All joints, (except the ones with screws) have been glued. The dimensions of the samples were selected so that sample remains relatively compact after the experiment and we didn't want it to burn away completely so that it was not necessary to extinguish the sample after the experiment. This way, we could measure the evaluation criteria called discontinuous criteria.

We achieved quite surprising results. All the joints reached worse values within the evaluation criteria against the control sample. These findings prove the following; we need to pay more attention to joints as such meaning that our hypothesis was correct; the conditions of the experiment (samples—their dimensions, intensity and the time of the thermal load), as well as other conditions of the experiment (including the evaluation criteria) had been chosen correctly. The results have confirmed not only that the control sample reached the best results, but we also noticed some minor differences between the various types of joints. The horizontal finger joint and the dovetail joint achieved the worst results. We assume that this is due to the fact that in the case of these two joints there is a direct heat transfer through the joint line, which is quite long even though the joint was glued. There were no major differences in the other joints. Vertical finger joint achieved the best parameters in all indicators. The obstacle to heat transfer was multiple indentations formed by glued wood.

The paper has brought a new perspective on wood joints even for fire protection purposes. The follow-up experiments should be more extensive though, using large-scale tests which don't include so many types of joints. Thermal and mechanical load of such joints combined in one test would be ideal. More scientific and experimental tests will bring new knowledge for the safe use of wood in practice.

It is only up to architects, engineers and investors to incorporate new knowledge of fire-technical characteristics of wood or wood-based products so that they can build safer constructions and use the whole range of wood and wood-based materials to our advantage in the construction industry.